MAKING BETTER CHOICES

MAKING BETTER CHOICES

Design, Decisions, and Democracy

Charles E. Phelps

AND

Guru Madhavan

OXFORD
UNIVERSITY PRESS

OXFORD
UNIVERSITY PRESS

Oxford University Press is a department of the University of Oxford. It furthers
the University's objective of excellence in research, scholarship, and education
by publishing worldwide. Oxford is a registered trade mark of Oxford University
Press in the UK and certain other countries.

Published in the United States of America by Oxford University Press
198 Madison Avenue, New York, NY 10016, United States of America.

Library of Congress Cataloging-in-Publication Data
Names: Phelps, Charles E., author. | Madhavan, Guru, author.
Title: Making better choices : design, decisions, and democracy /
Charles E. Phelps, Guru Madhavan.
Description: New York : Oxford University Press, 2021. |
Includes bibliographical references and index.
Identifiers: LCCN 2020047136 (print) | LCCN 2020047137 (ebook) |
ISBN 9780190871147 (hardback) | ISBN 9780190871161 (epub) |
ISBN 9780190871178
Subjects: LCSH: Decision making. | Social choice.
Classification: LCC BF448.M345 2021 (print) | LCC BF448 (ebook) |
DDC 153.8/3—dc23
LC record available at https://lccn.loc.gov/2020047136
LC ebook record available at https://lccn.loc.gov/2020047137

DOI: 10.1093/oso/9780190871147.001.0001

1 3 5 7 9 8 6 4 2

Printed by Integrated Books International, United States of America

In memory of
Kenneth Arrow and Michel Balinski

CONTENTS

CONTENTS

PROLOGUE

From One to Many

This is a book about how we collectively make decisions and about the tools we use to reach those decisions. The tools we use may be as important as the decisions themselves.

We make joint decisions out of necessity because the choices we make affect each other. Each decision we take has a consequence— sometimes perhaps very little, as in a private decision on what to have for breakfast, or perhaps a bit more, on whether to exercise or not. Some decisions involve jointly planning for dinner or vacation, or entering into a business partnership. Some decisions follow rules, laws, and constitutions. These decisions involve and affect large groups of people and organizations. The consequences are also greater and could last longer.

The problem of converting preferences of individuals into a collective decision has been pursued by a small academic field. The so-called social choice analysis tracks back at least to the 13th century, perhaps much earlier. The topic gained prominence in the late 18th century as the French Academy of Sciences sought reliable methods to elect its new members. Imagine such a debate between some famous people in science and public affairs: Laplace, Borda, Condorcet, and Napoleon Bonaparte.

Social choice analysis, though, mainly looks at what effects voting rules can have when used properly and manipulatively. This has implications in all sorts of scenarios in life. Consider the following: social clubs and professional groups electing new members; ranking of research proposals to be funded by taxpayer money; choosing space and telescope missions; exploring sites for major science and engineering projects such as the Large Hadron Collider in Europe; homeowners' associations choosing their directors and setting their annual fees and property rules; publicly traded corporations selecting their boards of directors with annual votes; and flying and sailing clubs voting about their fleet composition. And of course, we hold local, state, and national elections. Similarly, papal elections for the Church of Rome have become international events, as people wait for the telltale white smoke signaling a successful vote at the Vatican. In another context, in places like California, certain water districts give their members voting power in proportion to the acreage of land they own. In concept, one large landowner can control the vote to impose a tax on everybody living in the area to support the water supply. Making the most prudent decision remains a challenge throughout our lives.

Social choice analyses are also very limited in scope. They study how groups can choose among a given set of options. The menu is set. Groups are given a list of options. Groups vote. End of story. The issue of how the menu is set is not addressed.

Setting the menu is the realm of systems analyses, an approach most commonly associated with engineering design. While appreciated in concept and practice throughout the history of civilization, systems analyses came to prominence during the height of the 20th century war efforts when much more seemed to be at stake, including military superiority, space missions, and Cold War competition. Systems analyses is common in urban planning, waste management, environmental considerations, supply chain, and logistics, and it should ideally be even more common given the consequences of our work and how we are frequently unable—and unwilling—to track them. In simplest terms, systems analyses enable us to deliberately and consciously bring together factors that can and do matter in a situation. It's about recognizing that any decision has multiple criteria to satisfy—it's never one thing to vote on, and it's never one single answer. However, that is exactly how we have designed our commonly used voting systems to be: to come to a single conclusion without any consideration of expression, experience, and enjoyment. Engineers are well versed in certain aspects of design on how to integrate these key dimensions of life into the products and services they are building.

But until now, the task of systems design and the task of choice analysis within a system have been largely separated, particularly in settings where groups need to make decisions. In our decade-long collaboration, we have studied many decision-making processes. We have come to realize, from our perspectives as an economist and an engineer respectively, that the rules influencing how decisions are made seldom receive sufficient attention. But we have also learned in this project that organizational bylaws are imponderably vague—perhaps deliberately opaque—on some key steps in the decision processes. Further, these rules often give certain individuals vast power to control the process through designation of key groups and their own process rules. Thus, we were convinced that much work needs to be done on shaping the rules— or at least in making people aware of them—that guide social decisions. These rules are, in effect, the shadow decisions behind the real decisions.

We hope that this book will reinforce why we need better systems design and analyses given the consequences of our decisions. It is also about carefully thinking about the values of the choices we make, whether they be in a small meeting of individuals in your local association or community or in a national election. It will illuminate the differences between sincere voting and strategic behavior to defeat an opponent in voting, the latter being quite common. The book will also review different voting systems, what their original intents were, and what their deficits are. In trying to bring all these topics together and more, we realized that the book is in essence an outcome of the arranged marriage between social choice and systems engineering. The more one begins to explore the aspects of social choice and systems engineering, the more one realizes how much they have in common, and how much more they can offer if unified.

Chapter 1

Systems of Complexity

Around the world, men and women participate in track and field
events. They compete for glory, every year, every summer, every
winter. With commercial endorsements for high-profile performers,
financial gain remains a motivator to perform well.

The first Olympic decathlon event was held at Stockholm in
1912. The winner was Jim Thorpe, an American and the first Native
American to win an Olympic contest. King Gustav V of Sweden told
him after his win, "You, sir, are the greatest athlete in the world."
This tag—the world's greatest athlete—has stuck on future decathlon winners. Bob Mathias twice won the Olympic decathlon
(the first at age 17) and rode that fame to a career in the U.S. House
of Representatives for four terms. Bob Richards won two U.S. decathlon championships, becoming the first person to appear on the
front of a Wheaties "Breakfast of Champions" box in 1958, followed
by Bruce Jenner in 1976 after his victory in the Olympic decathlon.

Combining the performances of multiple events (10 for the
men's decathlon, seven for the women's heptathlon) is a quin-
tessential situation involving many attributes. The International
Association of Athletic Federations (IAAF) uses a complex formula to do this. The alleged goal is to make each event contribute
equally to the total score in terms of some standard of "fairness"

Making Better Choices. Charles E. Phelps and Guru Madhavan, Oxford University Press (2021). © Oxford
University Press. DOI: 10.1093/oso/9780190871147.003.0001

that remains unspecified. As we'll see, the IAAF systems do not come close to achieving this goal. Let's dive into the IAAF scoring formulas for a bit.

For the four running events (where T is time in seconds and "less" is better):

$$\text{Points} = a(b - T)^c$$

where a, b, and c are specific to each event.

For the six field events (where D is distance in meters and "more" is better), which include three throwing events, two jumps, and pole vaulting:

$$\text{Points} = a(D - b)^c$$

The formulas begin with the somewhat arbitrary choices on the b and c parameters. The b values are "minimum acceptable performances." For the throwing events, these minimum values are 6.5, 5.4, and 6.7 percent of the comparable world records. The leaping events—long jump, high jump, and pole vault—are next lowest (24.5, 30.6, and 16.2 percent of comparable world records). Then come the distance race of 1,500 meters and the 110-meter hurdle (42.9 and 44.9 percent of the world record), and then the 100- and 400-meter races (53.2 and 52.7 percent). These choices have a direct effect on scoring, since smaller minimum values (as in the throwing events) make it harder to gain points, and conversely. They have the effect of biasing the scoring toward sprinters and away from larger, stronger, and perhaps slower participants.

Another powerful effect comes from the c coefficients (power multipliers), which give a direct reading on the apparent importance of the event to the IAAF. The four running events have power multipliers ranging from 1.92 (110-meter hurdles) to 1.81 (100- and 200-meter sprints). The three leaping events have power numbers of 1.42 (high jump) to 1.35 (pole vault). The three "throwing" events have power numbers of 1.10 (discus throw) to 1.05 (shot put). These closely correspond in effect on scoring to the b multipliers, favoring speedy sprinters and disfavoring stronger and larger athletes.

After the IAAF agrees on the b and c values, they pick a scores so that a prespecified strong performance nets 1,000 points. IAAF historical documents say that the intent in the 1912 Stockholm Olympics was to award 1,000 points in each event for the Olympic record (typically less than the world record, which would produce about 1,200 points), but the formula has been modified six times since then.

In the real IAAF process, the scoring system favors the fast over the strong, but it is very hard to decipher how exactly. The three parameters (a, b, and c) interact in a complex way and show what's important for each event. The decathlon and heptathlon scoring methods may seem arbitrary and capricious and may disfavor some events, such as throwing, over others, such as running. This creates some selection pressure on individuals who can compete effectively in the men's decathlon or women's heptathlon. Again, lighter and faster athletes have an advantage over stronger athletes. That is, it's not a pure competition where each event has an equal effect on the outcomes.

Two issues emerge. First, the ways in which multiple values are combined affects the final results. Second, and perhaps more important, these grading rules can and do change the nature of possible choices. The rules we create influence what's considered valuable and affect the choices presented to us.

.·.

Our collective decisions usually revolve around three basic factors: (1) how many people are affected by the decision; (2) how many people make the decision; and (3) how important the decision is both to the makers and recipients of the decision.

Some decisions have one decider with one or two people affected, as in deciding what to have for lunch today. This is of hardly any consequence to society. A similar decision can have higher consequences, as in what car to buy, or of greater consequence, as in deciding on surgery versus hospice for cancer care. Another example of a decision that is made by one decider with at least two people affected, and that is of great consequence, is filing for divorce. Couples on the verge of a divorce usually don't take votes on such matters.

Some types of decisions have—or at least appear to have—one decider but affect many people, as in decisions of a government to go to war, to shut down its operations, or to declare a national emergency. These decisions affect a wide range of people, perhaps an entire country or continent. Arguably, decisions made by a few individuals at the political helm of a few organizations and nations affected the entire world in the COVID-19 pandemic.

Such decisions (set aside totalitarian regimes) are seldom made by one person. The U.S. House of Representatives has sole jurisdiction to initiate tax bills. The U.S. Senate ratifies presidential, judicial, and cabinet appointments. The senators must also ratify U.S. participation in international treaties and in declarations of war. The ultimate decision authority remains with the people, who, through established procedures, elect the leaders of cities, states, and nations. So even the apparently significant decision-making power of the U.S. president is tempered, at least in concept, by the power of the vote. If people actually employed their full right to vote, elections of the U.S. president would both involve and influence the entire country.

In many types of decisions, the number of deciders and the number of affected people are both small. Elections for mayor, town council, sheriff, and judge are examples. They expand to the level of counties and states as well as many other forms of government, such as those that support water supply, education, and medical care. The same idea applies to the private sector.

Certain decisions involve an intermediate number (several hundred) of decision-makers in situations where millions (sometimes billions) of people are affected. Conversely, some decisions are made by groups of people but affect very few. The Supreme Court of the United States has nine voters, but on certain concerns, their decisions could affect everyone in the country (about 330 million in 2020). The College of Cardinals consists of 222 members, but only those under age 80 are eligible to vote for a new pope, and the number of electors is limited to 120. Their vote affects the estimated 1.2 billion members of the Roman Catholic Church around the world, if not more. The state of California has about 25 million eligible voters out of a total population of about 40 million, and while many of their votes have influence only within California, on policies for auto emissions and minimum wage the effects of their decisions can ripple throughout the United States. The result of the recent elections in the United States and the United Kingdom has had an influence on the entire world. In India, recent elections for national governance involved 900 million registered voters (out of

1.3 billion people), the largest democratic exercise in known space. Each of these scenarios requires agreement by the participants on the decision rules. However, a key point of this book is that people seldom think about the rules behind their decisions, even though these rules can and do have important effects on their own and other people's lives.

There are many predictable components in decision-making. Sometimes they get rolled into moments that cannot afford time for thinking, as in encountering a grizzly bear on a mountain trail. We have enough experience with fight-or-flight situations; these are instinctive and evolutionary. And paradoxically, we do not have much experiences in situations we ourselves have created: when decision-making needs to involve multiple people, various steps, numerous discussions, arriving at an agreement when possible, and launching the decision.

Many decisions first involve a process to whittle down a larger list of options to a manageable set. This step is ubiquitous and often strongly influences the outcome. Many organizations and corporations do this using committees of individuals. Sometimes these committees are permanent (with a rotating cast of characters who serve)—for example, a committee tasked with having ready a list of candidates at any needed time to fill vacancies on a board. Sometimes the committee is ad hoc and need based, as in selecting a new leader for an organization. The selection process and the voting rules to choose the final list of candidates are usually well specified and available for public view. In contrast, earlier steps have greater mystery. These rules are often not written down, and when they are written down they are available only to a select group of insiders. From college admissions to competitive awards, the mechanisms and decisions for creating a list of candidates to be voted on are likely to be less obvious and probably even obscure.

A high-profile example makes the point: the process of selecting winners for the Nobel Prize in Economics, formally the Sveriges Riksbank Prize in Economic Sciences in Memory of Alfred Nobel. The point here is not to say that the process is wrong or very secretive but rather that the obscurity of the process overrides its transparency. The members of the prize committee are elected by the Royal Swedish Academy of Sciences for three-year terms, but the details about how those committee members are chosen remain obscure to the public. The committee seeks nominations, typically from about 3,000 people. This pool includes former prize winners, members of the Royal Swedish Academy, and professors from around the world who are chosen by the committee. From the list of nominees, typically 250 to 300 in any year, the committee then determines a preliminary list of candidates in a secret process and seeks specialized evaluations of their work.

Next comes the majority vote. Even this step of the process is obscure. On exactly what attribute do they vote? It cannot be true that they choose the winner from a long list, because (as will be discussed later) majority voting does not always determine a winner if three or more candidates exist. Plus, the committee has never failed to award a prize in any given year. So, they might choose among two or three finalists, except that sometimes the prize is awarded jointly to several individuals, dividing the funds between them. The vote might consist of two proposals from the committee, one of which might involve multiple participants. Or perhaps the vote is a "rubber stamp" vote to approve the committee's recommendations. Whatever process they use—and this is not just with the Nobel prize selections—we know that (a) it is not publicly described in detail; (b) it apparently always produces a winner; and (c) the winner might be one person, or two, or occasionally three. In the

50 years of awarding this prize, half of the years have had multiple winners; in six years, three winners have shared the prize.

Again, the basic point here is not about the secrecy of the selection process (which is understandable) but more about the mysterious winnowing process that underpins this award and those in many competitive pursuits.

The rules (or lack thereof) surrounding our selection processes can and do affect the diversity and inclusion in society, the very essence and engine of democracy. This is well known and often acknowledged in theory but not in practice.

The importance of the decisions that are created by collective choice is a product of how many people are affected and the impact on well-being. Decisions of high consequence that affect large numbers of people deserve special attention. Decisions are also a function of who can make decisions and if those rights and privileges are clearly specified. The U.S. Constitution specifies those rights. Every country, state, county, and municipality in the world probably has some comparable document. But these statements are often abstract and can easily be interpreted differently, as can be seen in debates concerning the powers of the presidency or legislature in the United States. Constitutions regularly specify judicial systems, for which another set of pivotal decision rights is laid out. In the United States, the ultimate "decider" on consequential matters of law is the Supreme Court. In totalitarian societies, a dictator often holds final decision rights over almost everything, despite the extent to which the nation's official name tries to otherwise persuade (e.g., "The Democratic People's Republic of X").

Next, when decision-makers "decide," what should they decide? The best choices, we submit, will not necessarily involve choosing among candidates but rather choosing what is valuable—and by how much and to whom—and also finding how well these candidates perform against what we value. Many decision support systems have become very popular and easier to use. The value of these aids has become increasingly apparent as we learn more about the foibles and (sometimes) major defects of intuitive human abilities to make decisions. We struggle with probabilities. We use estimates incorrectly, even inappropriately. Our judgment is affected by how questions are framed. Decision support systems play a good role in helping us recognize and minimize cognitive errors.

Three key issues have emerged. Who gets to decide? What are they deciding? What rules and choice mechanisms do they use? On these topics, many mathematically dense publications on social choice only offer "caution." These are well intentioned but do not provide useful advice. In *Making Better Choices*, we'll address these questions and also consider the human factors of those choice mechanisms. How do they work? How friendly are they? How expressive are they? How meaningful are they across contexts? Throughout this book, our goal is to focus not on the rules of voting but on the rules of discourse—how to design better decision mechanisms in democracy.

Before we launch into more detail, let's consider a practical insight from systems engineering: We must make tradeoffs, which in turn requires knowing the relative importance of our priorities and what we consider valuable.

In the late 1970s, all high-end bicycles (including those used on the Tour de France) had steel frames. Fine distinctions were drawn between various types of steel tubes that best served the triple goals of competitive bicyclists—low air drag (smaller diameter), low weight (thinner walls), yet stiff as possible to avoid frame wobble. The relevant design parameters were choice of steel alloy, tube diameter, and wall thickness. This frontier of possibilities led bike makers to different choices, sometimes small and irrelevant. Climbing steep mountain passes in the Alps and then racing down the other side at 60 mph created the greatest stiffness challenges: Insufficiently stiff bikes cost a lot of energy on the climb and then wobbled dangerously on the descent.

Then along came engineer Gary Klein, who wondered while a student at Massachusetts Institute of Technology if a way existed to break the paradigm. For a design class, Klein began experimenting with aluminum tubing of substantially greater diameter than existing steel-tubed bikes. The larger-diameter tubes were much lighter and much stiffer but had greater air resistance. A tube's stiffness relates directly to the fourth power of its diameter. When you double a tube's diameter, it becomes 16 times stiffer, which greatly improves the bicycle's performance in competitive races.

Klein revolutionized bicycle construction.[1] His design was vastly better than any other bike on the market at the time. Tungsten-inert-gas (TIG)-welded aluminum tubing has been replaced by carbon fiber, yet another technology that pushed the performance frontier even farther. Only cost stands in the way of using carbon nanotubes as the next-generation foundational material. This could

---

1. One of us (Phelps) rode a Klein Stage bike across the United States from Oregon to Maine, averaging 100 miles a day for five weeks. Not only was it faster and more efficient than steel-framed bikes, but it was also remarkably more comfortable to ride.

allow thinner tubing, hence less air resistance, but trade-offs would remain. In the long run, Klein's key idea was not to use aluminum but rather to make fatter-tubed bicycles, which rapidly evolved into oval rather than round tubes, and other materials beyond aluminum to do this. But even with this improvement, trade-offs remain as an essential part of bicycle design.

Automobile design faces the same conundrum. Car design always has competing goals—say, fuel efficiency (miles per gallon), acceleration capability (torque), and emissions reduction. Limited to the technology of combustion engines (gasoline or diesel), you can't have all three. Some have tried (by deception), but the iron laws of physics—and the practical necessities of engineering design—require tradeoffs. The market recognized this with a one-third drop in the share price of the firm within days of its emissions fraud becoming public. You can't improve on all of these goals at the same time. Yet (like with Klein's aluminum tubing) automakers have found a way to improve on all three dimensions at the same time by changing the paradigm from internal combustion to stored electricity—batteries. But this adds three new dimensions to the problem—weight, range, and cost. The concern with electric cars to date has been their limited range. Someday, a new battery technology will alter that equation, or hydrogen-powered fuel cells will come into play. Then a new constraint will emerge, and new trade-offs will need to be made.

An enduring insight from the practices of systems engineering is that you can't solve all of your problems at once; tradeoffs are essential. Consider this adage: If quality, speed of completion, and cost are three variables, you get only two of them. The skill, even achievement, becomes how to arrive at good-faith tradeoffs, and this is where the practice of systems engineering tries to ensure reasonable and maximal levels of quality, speed, and cost.

There are many denominations of systems engineering according to the industry and contexts, but variations of the following themes are put into practice: development of a strong conceptual understanding of the problem and solution space, generation of requirements (options) for the desired solutions, periodic verification (as in quality assurance) and validation (as in ensuring safety) through multiple rounds of testing, and iterative improvements in the processes, services, and products. Although simple in description, this kind of engineering has to deal with complexities at every level: business objectives, human psychology, market needs, political forces, and time constraints. Even systems are of different kinds: simple systems with a single path to a single solution; complicated systems with multiple paths to a single solution; and complex systems with multiple paths to multiple solutions. In complex systems, the properties and character of the phenomena change frequently, fundamentally challenging assumptions and established processes that work for well-understood systems. Systems engineering in this case becomes a process of disciplined learning and unlearning. The approaches may not necessarily yield a pinpointed solution per se but can offer an insight on the nature of compromises to be made, which is true of scenarios in making both decisions and policy. But the hard practical fact remains: Life is replete with tradeoffs.

## Chapter 2

Value and Values

What car should one next purchase? This is a common enough financial decision faced by individuals. One could think about similar problems, such as which job offer one should accept or which house a family should purchase. A typical approach for joint decision-making would have the relevant participants in the family set out a list of potential choices—a list of car models, or at least brand names of interest. Then the family would talk them through, whittle down the list, take some test drives, haggle about the prices with dealers, and then make a choice. The decision-maker may be an individual, a couple, or several members of the family, but the unit of choice is the vehicle. The choice is made among specific candidates—cars, SUVs, trucks, and so forth.

An alternative process exists that may be more work but may also lead to more informed choices. Suppose instead of selecting among specific vehicles, the family members write down a list of what features of vehicles are important to them—that is, describing vehicles with attributes. They then figure out how much each one of these features matters relative to the others, and then they evaluate a range of vehicles to see how well they meet those criteria. The trick becomes how to combine them in a meaningful way. For a family with several children, some of driving age and some younger,

Making Better Choices. Charles E. Phelps and Guru Madhavan, Oxford University Press (2021). © Oxford University Press. DOI: 10.1093/oso/9780190871147.003.0002

a sample list of important attributes might include the following: It should have decent seating capacity (say, at least five people for a large family), outstanding safety features, terrific fuel efficiency, large luggage capacity, rapid acceleration, good braking ability, and a built-in entertainment system for children, and it should be able to withstand crashes well, all at minimal cost (think: teenage drivers). This particular family may be thinking they are in the market for a minivan or possibly an SUV or large sedan. If they just think minivans, they may miss some important options.

A different family—say, a relatively young couple with no children (and none on the way)—might value a different set of things. They may want a car that seats two, has great fuel efficiency, is small and easy to handle in traffic, has a sporty feel, has good acceleration and traction, has a great audio system, and has low repair costs with an extended warranty. We can think of many other prototypical families, but the obvious point is that different families will value different things and will almost certainly be shopping in different segments of the automobile market. Is there a way to better assist them in their decision-making process?

The typical way for humans to think through choices is more intuitive. The process usually involves taking the candidate cars out for a short test drive, talking to friends, and reading online reviews or watching videos. Then, they make a choice from the cars they can afford. What's really happening here is that our brains—mostly subconsciously—are integrating a lot of information and coming to an intuitive conclusion. Alas, as is well understood, we humans do this quite imperfectly. That's where decision-support systems can help. By leaving complex data and computations to computers, these systems capitalize on what humans are best at: determining what "value" means.

Before venturing into the world of these decision-support models, let's explore briefly some of the more compelling reasons to use them, as viewed through the lens of human behavior and drawing principally from the scientific insights of Daniel Kahneman, Amos Tversky, and Richard Thaler and their collaborators.

The law of small numbers. People too often see patterns or draw conclusions from overly small samples of data. Try this experiment yourself. Flip a coin four times and record how often it comes up all heads or all tails. On average, this occurs one out of every eight times. If you do the same experiment with groups of seven coin flips, they will all be the same about one time out of 64. Small samples can lead to erroneous conclusions, as in "this coin is rigged." This also leads to overconfidence in what one "knows" and leads to overly rapid assignment of causation when patterns occur. Most mystical omens have their root in this—that is, some X causes bad luck, be it breaking a mirror, walking under a ladder, seeing a black cat, or drawing a specific card out of a deck.

Anchoring. Irrelevant or misleading numbers can affect people's estimates of factually knowable values. A good example comes from an experiment at the San Francisco Exploratorium. Participants are asked: Is the height of the tallest redwood tree more or less than 1,200 feet? Then they are asked to guess the actual height of the tallest known redwood. The average answer is 844 feet. However, if the first question uses 180 feet instead of 1,200 feet, the average answer is 282 feet.

Framing. The ways in which questions are posed or even the ways hard data are presented affect decisions. In an experiment at Harvard University, medical center patients, graduate students, and physicians (over 1,100 in total) were asked which of two therapies they would prefer to treat lung cancer: radiation or surgery. The surgical risk was described either as "a 10 percent chance of dying" or "a 90 percent probability of survival." The decisions heavily tilted toward surgery when the "probability of survival" phrase was used and much more toward radiation when the "chance of dying" phrase was used.

Availability. When examples of a specific idea readily come to mind, people tend to generalize their experiences, thinking they are more frequent. Say that you or a loved one were to become a victim of medical malpractice. That experience will increase your belief that malpractice happens more frequently—more so than if we see news stories about the same event happening to people we don't know. Stories about sex scandals in Hollywood, business, or politics will lead you to believe that these events happen more than in other areas of life, whereas similar stories about college professors whom you don't know will have a lesser (or no) effect on your beliefs about sexual harassment in academia, because the names of the people are less "available."

Status quo bias. Defending the status quo seems to have innate evolutionary roots. In disputes over territorial control, the "resident owner" wins almost all contests from interlopers in all animal species. Human preferences for the status quo have repeatedly been shown. For example, when people are asked to choose between two alternatives, one representing the status quo and the other a change, strong preferences are shown for the status quo, no matter which option carries that label.

Emotional risk estimation. Vivid and scary events heighten people's perceptions about the overall risk they create. People estimate strokes and accidents as about equally deadly overall, whereas strokes cause twice as many deaths as all accidents combined. Death by accident was estimated to be 300 times more likely than death from diabetes, when in fact diabetes causes four times more deaths than accidents. Tornados are estimated to cause more deaths than asthma, even though 20 times more people die each year from asthma than tornados. "Going viral" on social media is likely to exacerbate this effect. A small event that "goes viral" becomes an event in itself, magnifying by far the importance or frequency of the original event.

Representativeness. People often conflate and routinely mix up two very different types of statistics. Among those who drink any alcohol, the "beverage of choice" differs by educational attainment. Among college graduates, 44 percent prefer wine, 35 percent beer, and 18 percent liquor. Among those with high school or less educational attainment, 52 percent prefer beer, 24 percent liquor, and 21 percent wine. Representativeness would lead one to conclude that a person seen drinking beer has less education. However, overall drinking rates also differ by educational attainment. Eighty percent of college graduates drink alcohol, while only 52 percent of those with a high-school education or less do. The "high school or less" group and the "college or above" group have approximately the same number of people in them, so the person you see drinking beer is actually more likely to be a college graduate (or higher) than a high-school graduate (or less). This is a simple error involving Bayes' theorem, a well-known statistical relationship.

Less is more. Sometimes people try to make too many things fit a story. In an oft-repeated experiment, psychologists created a profile of "Linda" as a classic social activist, like a prototypical radical student from Berkeley in the 1970s. They then showed participants a list of potential occupations she might have or activities she might be in, including schoolteacher, social worker, bank teller, active in the feminist movement, insurance sales, and (the key to their experiment) "bank teller *and* active in the feminist movement." The participants were asked which of these scenarios had the highest probability of being true. Illogically, more of them chose "bank teller *and* active in the feminist movement" than "bank teller" alone. This cannot possibly be true logically: One is a subset of the other and hence cannot occur with higher probability. This is yet another example of how humans intuitively struggle with statistics.

Next, let's look at some issues with estimating probabilities.

Mixing probabilities together. Lower probabilities (say, under 20 percent) are interpreted as "won't happen." Large probabilities (say, over 80 percent) are interpreted as "will happen." Probabilities closer to 50 percent are interpreted as "might happen."

"Vivid outcomes," or confusion with the probability of happening. Think about buying lottery tickets, an activity that a colleague described as "a tax on people who don't understand probability theory." On average, one will lose money playing the lottery. But people behave as if the chances of winning the lottery are much higher than they really are, probably because winning the lottery is so desirable.

"Vivid probabilities," or how data presentation influences statistics.
When probabilistic information contains actual number of
affected people (rather than abstract percentages), people
react more strongly to the numbers. This has been called the
"frequency format" or the "denominator effect." People visu-
alize the numerator and ignore the denominator when they
are given numbers instead of ratios. If you say one person in
5,000 will die from *X*, people will view it with greater alarm
than if you say the risk of dying from this event is 0.02 percent. We can vividly imagine that "one person." If disease *X*
kills 1,286 people out of 10,000, people will judge it as more
dangerous than saying that it kills 24.14 percent of the population. The "vivid" number of 1,286 dead people heightens
the perceived risk, and people mistakenly give the wrong answer quite commonly, even though it is easy to see with a
little mental math or a calculator that the second risk (24.14
percent) is almost twice as large as the first (12.86 percent).
Personally experiencing probabilities. People tend to understate
the probabilities of events that they have never personally experienced. Even among those living in Southern California,
most have never experienced a really large, devastating earth-
quake. Many of them, however, have experienced a number
of relatively small temblors. They discount the probability of
"the big one" happening in their lifetime, and this may even
lead them to act as if it will never happen.

At its core, decision-making is about perspective-taking. How
one pursues this exercise is a function of how many variables
need to be dealt with, along with all the fallacies, biases, and
probabilities. Economists study models where the *level* of wealth
or income creates some happiness, labeled as "utility." The

development of prospect theory proposed that *changes* in wealth or income can create utility. This theory explains and magnifies a well-known distinction that economists mostly overlook—the willingness to pay (WTP) to avoid an externally imposed risk such as illness, automobile collisions, and fires versus the willingness to take on new risks, or willingness to accept (WTA). Economists normally assume that WTP and WTA are nearly identical. In fact, when they are directly compared, WTA exceeds WTP by factors of five to seven. People might be willing to pay large sums, perhaps even all of their fortune, in ransom to a kidnapper to avoid being killed. They could well require an infinite amount to voluntarily accept the same risk.

As a less violent example, consider this simple experiment. As students enter the classroom, half of them, randomly chosen, receive a coffee mug, emblazoned with their college logo. As the class proceeds, students are asked to inspect the mugs and then to write down how much they would be willing to pay to have one. The students who received the mugs were asked how much they would need to be paid to give it up. In an ideal sense, the amounts should be somewhat equivalent, give or take. In fact, as has been shown in widely replicated versions of this same experiment, the WTA is about twice the WTP. Once students "own" the object, it becomes much more valuable to them.

The same logic is responsible for the famous "Not In My Back Yard" (NIMBY) phenomenon. Everybody wants excellent cellphone or internet service, but nobody wants the cell towers in their neighborhood, let alone on their own property. Everybody assumes that a stable electricity supply is vital to our society, but nobody wants high-voltage lines near their home. Most people want the convenience of air travel, but nobody wants an airport to be sited near their home. This can be seen in more extreme forms of thinking such as

"Build Absolutely Nothing Anywhere Near Anything" (BANANA) or "Some Other Bugger's Back Yard" (SOBBY).

Moreover, there are institutionalized and systematic biases that become critical in the context of decision-support systems that may help illuminate—or at least properly account for and potentially reduce—racial, ethnic, gender, and other undesirable biases. Some prominent examples include:

Selection bias. Symphony orchestras regularly audition for new performers. It became well known that women had greater difficulty in winning these auditions in almost all instrumental categories of classical music. Obviously, in vocal auditions, the voice range defines "the part" and hence whether the candidates are male or female, but not so in instrumental auditions, in which the gender should be irrelevant. Recognizing this potential bias, orchestras have commonly shifted to blind auditions, wherein performers walk to the stage and perform visually shielded from the judges. This partly addressed the problem, but not entirely. Judges commonly heard a difference when women walked onto stage (even hidden from view) in high-heeled shoes, so the bias against women persisted to some degree. Women were then advised to wear soft-soled flat shoes into auditions to remove this signal. Even then, some signals still persisted. A leading oboe teacher notes that some can discern whether an oboist is male or female just by their breath sounds, since women are generally smaller than men and hence have less lung capacity.

Racial stereotyping. The cities of Boston and Chicago have had a long and checkered history of racial problems, with a high degree of racial segregation that is well understood in local business and industry. Researchers created fictitious résumés and sent them in response to a variety of published help-wanted ads to assess the degree of racial bias. The résumés were created from actual postings in local labor market websites, synthesized to create realistic combinations of education and experience. The résumés created from actual Boston-area websites were used in Chicago, and vice versa, substituting high-school names appropriately, for job listings in sales, administrative support, clerical services, and customer service. Using birth certificate data, researchers identified and randomly assigned names commonly associated with African American families (e.g., Lashika and Jamal) and those more commonly associated with White families (e.g., Emily and Greg). White-sounding names received 50 percent more responses for a job interview than Black-sounding names, confirming differential treatment in the job markets. This type of research has now been replicated broadly in numerous counties relating to employment, housing, and other similar endeavors, and differential treatment has been demonstrated to exist regarding racial minority status, sex and gender roles, and even personal appearance traits such as obesity. It has been documented in every continent in the world (except Antarctica). It seems to be a ubiquitous human trait, although it has not yet been observed in penguins.

Inequities in medical care. Another widely understood form of racial bias is literally a matter of life and death. African American women and men are more likely to die from preventable diseases and conditions than White women and men in comparable situations (e.g., income, insurance coverage, urban living). This finding has been repeated over and over again in

terms of heart disease, cancer treatment, and—astonishingly
in a highly developed country such as the United States—
mother's death in childbirth. In the United States, African
American and Native American mothers are three or more
times likely to die in childbirth than White, Hispanic, and
Asian American mothers. Some of these excess deaths come
from the childbirth event itself (hemorrhage, pulmonary em-
bolism, infection, and sepsis) but others stem from underlying health conditions (hypertension, cardiac disease) that
remain untreated in the high-risk groups.

In summary, humans make numerous decision errors that
compound themselves. Many of the underlying fallacies relate to
the ways people estimate probabilities, and then how they (mis)
apply them in decisions. Both of these types of errors come readily
into play with intuitive decision-making for complex problems.
Some involve racial, gender, and ethnic bias, either deliberate or
subconscious. Can we avoid them? Yes, and systems engineering
approaches offer pathways to bypass cognitive errors.

Systems engineering tools, in their generic sense for use in decision
support, provide an organized format to analyze alternatives. In general, these are multiple-criteria models enabling a broader look at
the scenario. They often include choosing the attributes that matter,
deciding how much weight should be placed on each of them to
express emphasis, organizing data for these attributes, and scoring
the choices for comparisons and discussion. These are decision-
support systems, not decision-making systems. This process has
humans doing what they do best—specifying what has value and by
how much—and letting computers do what they do best. Scoring

the candidates on objective measures of value often involves complex mathematical calculations that blend measured performances and probability estimates.

In our previous work, we confronted complicated analysis to estimate the key attributes that vaccines might have. These included such things as lives saved (involving disease-specific mortality rates, vaccine uptake rates, and efficacy), life-years saved (blending the previous information with life-expectancy tables and distributions of the age of specific populations), and even quality-adjusted life years (adding measures of quality of life in different illness states). There were also computations regarding medical expenditures that might be avoided and other complex combinations of data. Our analyses also included subjective attributes that could not be measured from real-world data. For these, our software users must "score" the alternatives subjectively, according to the needs of their contexts. Some examples include vaccine benefits to certain populations of interest, assists in foreign policy negotiations, reductions in the threat of bioterrorism, reduced reliance on the "cold chain" (refrigeration from production to end-use), and fit with local immunization programs. Some of these can be elicited through population or expert opinions, but others are in the domain of the decision-makers.

The multi-criteria models then combine these parameters and preferences into a unified score for comparison against available choices. In the simplest model, the value of candidate j is something like this:

$$V_j = w_1 Z_{j1} + w_2 Z_{j2} + \ldots + w_k Z_{jk} = \sum_k w_k Z_{jk}$$

where the weights add up to 1.0 (i.e., $\sum_k w_k = 1$) and the Z_{jk} values provide a score for each candidate (j) on each attribute (k). To do this right, we need two things: (1) a good process to elicit the weights and subjective preferences from decision-makers and (2) a way to convert all of the differently measured attributes into a common scale, so that it makes sense to add them up to place them in a common measurement logic such as scores on a 100-point scale. Often, the Z_{jk} scores are rescaled versions of the directly measured attributes, which we will call the X_{jk}. Commonly, that rescaling involves creating a "measuring stick" for each attribute that scales everything into a common 0-to-100 interval.

Multi-criteria models do these tasks in different ways. Perhaps the most common models use the range of actual performances of the candidates to create "measuring sticks" for each dimension of value. They do this by using the best- and worst-case performances of the candidates across each attribute. This sets a scale and helps compare the candidates in a standard fashion.

Going back to the example of a family selecting a car, suppose we are assessing two of the car attributes that are measurable: fuel efficiency and ability to stop safely and rapidly. Miles per gallon (MPG) is a direct indicator of fuel efficiency, where "more is better." For ability to stop, the standard measure is distance from 60 mph to zero, measured in feet, where "less is better." Suppose that we also have a third, more subjective attribute, the car's classiness.

A common method for scaling MPG would find the best among available options (e.g., 56 MPG for a Toyota Prius) and the least favorable among available options (say, a Rolls-Royce Phantom, 14 MPG). The measuring stick for MPG then becomes the difference between the best and the worst: 42 MPG. So a car getting 33 MPG (like a Chevrolet Spark) is 19 MPG better than the worst and thus scores $19/42 = 0.452$ on the MPG scale—almost halfway between the worst and the best. The ability to stop rapidly might go (for the same set of vehicles) from 150 feet (worst) to 90 feet (best) for a scale of 60 feet. A car stopping in 100 feet would be 50 feet better than worst, so the score would be 50/60 = 0.833—over 80 percent of the way from worst to best.

For the classiness attribute, one could use scales like 0 to 10, where presumably the Rolls-Royce would score 10 out of 10 and a Jeep Wrangler might score, say, a 2 or 3 (it would do really well on off-road capability if that were an included attribute), but the scale would still be the previously defined 0-to-10 scale. These then get converted to the same 0-to-1 or 0-to-10 or whatever interval as the other scores. With this sort of scaling, one can add up the scores of each candidate across all of the attributes being valued.

We can also learn a good amount from the competitive selection practices that involve subjective sensations. Consider the ranking of wines—and we'll discuss a very high-profile example of the complexities in a later chapter where California wines were pitted against French wines in a head-to-head competition of the so-called Judgment of Paris. A standard approach has tasters score wines across different characteristics of each wine with a fixed number of points (say, 3, 5 or 10), with a total for a perfect wine adding up to a target score (say, 20 or 50). This approach gets the tasters to focus on different desirable "attributes" of the wines instead of having them give an overall summary score. In the latter, the tasters score each wine's different attributes (color, bouquet, flavor, overall taste, and so forth) and then add their own implicit weight to each attribute. Most wine-tasting processes eschew the summary score approach and concentrate on having tasters score on each attribute. Then the points assigned to each attribute define the weights. Here are three actual wine-scoring systems.

The Robert Parker system looks like this:

Color, appearance	0 to 5
Aroma and bouquet	0 to 15
Flavor	0 to 20
Overall and aging potential	0 to 10
Total possible points	50 (plus 50 base points)

The Australian wine rating uses this scale:

Sight, appearance	0 to 3
Smell, nose	0 to 7
Taste, palate	0 to 10
Total possible points	20

The University of California at Davis wine ratings use a more refined scale:

Appearance	0 to 2 points
Color	0 to 2 points
Aroma and bouquet	0 to 4 points
Acescence (volatile acetic acid)	0 to 2 points
Total acidity (tartness)	0 to 2 points
Sweetness	0 or 1 point
Body	0 or 1 point
Flavor	0 or 2 point
Astringency	0 to 2 points
General quality	0 to 2 points
Total possible points	20 points

The UC Davis scores can usefully be aggregated by sensory type to match the other systems:

Sight, appearance	0 to 4 points
Smell, nose	0 to 6 points
Taste, palate	0 to 8 points
General quality (overall)	0 to 2 points
Total possible	20 points

We can now see how much emphasis is given to each category of quality—visual, smell, taste, and the overall "gestalt" sense. Table 2.1 shows the allowed point totals on the left half and the implicit weights on the right half.

Although similar, these are not the same weighting systems, and the same set of scores from a wine judge (or aggregation from a panel of judges) could yield different results. The biggest difference comes with the allowance for overall quality, ranging

Table 2.1 THREE WINE-SCORING SYSTEMS

	Scoring points awarded			Implicit weights		
	Parker	UC Davis	Australian	Parker	UC Davis	Australian
Visual	5	4	3	0.1	0.2	0.15
Smell	15	6	7	0.3	0.3	0.35
Taste	20	8	10	0.4	0.4	0.5
Overall	10	2	0	0.2	0.1	0
Total	50	20	20	1	1	1

from 20 percent for the Robert Parker method to nothing for the Australian method. Of course, these weights make a difference, as they should, and of course, as the saying goes, there is no value in arguing about tastes. Individual tastes indeed define the scores.

Consider three fictitious wines—Ten Buck Chuck, Guru's Grapes, and Chateau Kayli.[1] We'll revise the way the wines are scored slightly to make them identical in concept to multi-criteria scores that normalize each attribute (here, on a 1-to-10 scale) and then apply the weights as revealed in the three scoring systems (shown in Table 2.1), scaled to total 20 points for a "perfect" wine. This replicates the 20-point scales of UC Davis and the Australian system, and it scales the Parker score so it also has a 20-point range (the normal 50-point range divided by 2.5).

In Table 2.2, the left half gives a single taster's scores on each component of the three wines (visual, smell, taste, and overall impression, each rated on a 1-to-10 scale), and the right half gives the final scores. The winner differs under each ranking, despite the same data (the judge's scores) being used in each case. The final score depends on how the attributes of the wines are weighted.

The Parker method winner (Guru's Grapes) is weak on visual quality (5 out of 10), but the Parker model places the least emphasis on that, and it scores well on taste and overall, which account for much of the Parker system weight. The UC Davis winner, Ten Buck Chuck, is strong on visual (10 out of 10), which gets the highest weight (20 percent) in the UC Davis system. The Australian system

1. "Ten Buck Chuck" is a take on what used to be called "Two Buck Chuck," the Charles Shaw wine series from California grocery store chain Trader Joe's, which now sells for $3.99 per bottle, and perhaps would best be called "Four Buck Chuck." Guru does not drink, and as a lifelong vegetarian is fond of grapes. Kayli (inspired by Hindu goddess Kali) is the name of the black female Standard Poodle in the Phelps household.

Table 2.2 THREE WINES, THREE RATINGS, THREE WINNERS

	Visual	Smell	Taste	Overall	Parker	UC Davis	Australian
Ten Buck Chuck	10	8	7	7	15.2	15.8	15.2
Guru's Grapes	5	7	9	8	15.6	15	15
Chateau Kayli	7	5	10	7	15.2	15.6	15.6

winner, Chateau Kayli, has the highest possible taste score (10 out of 10), which gets 50 percent of the weight in the Australian system.

So, which is the best wine? If you rely on rating systems like Parker, Australian, or UC Davis, you implicitly accept the weights that they place on each attribute. However, in standard use of multi-criteria models, you would apply your own weights to the attributes to create your own unique score.

When decisions have multiple factors and influences, and particularly when these involve probabilities and big consequences, having humans decide among final candidates may well be the less desirable choice. This requires that humans do a lot of complex integration in their heads. We don't do that very well. Multi-criteria decision models used in the practice of systems engineering provide good support for whole analyses and can help bypass many cognitive errors. However, the work required to create and use a multi-criteria model is not trivial.

How, then, should one decide when to use systems engineering–based methods for decision support instead of plain intuitive decision-making? Ironically, of course, that judgment needs to be made intuitively. Perhaps one way to sharpen our judgment in these situations is to make a series of intuitive decisions where the answer is "maybe yes, maybe no" in terms of using decision-support models, and then go through a modeling exercise. At the end, one can then compare the decisions to see which seems better, giving some sense about when it will be worth the extra effort.

Chapter 3

Decision Rules

The dean of a college has authorized a new faculty hire in the
Department of Metaphysical Postmodern Poetry. The department
chair appoints a search committee to evaluate candidates. They post
advertisements in the *Journal of Advanced Theoretical Postmodern
and Contemporary Metaphysical Poetry*, the leading journal in their
field, announcing the opening and inviting applications. They
call their friends in the field around the country at universities
producing doctoral candidates of appropriate quality to seek possible applicants. CVs flow in, along with letters of recommenda-
tion. All of these specify that candidate X is among the top three or
four candidates in their college or country coming out that year (or
over the past several years), and urge that candidate X be hired. The
search committee selects a handful of these to give job talks in the
department's seminar series. The faculty meet, discuss the choices,
and rank them, advising the chair to negotiate a deal with the top
choice, subject to the dean's approval, and failing that, go to the
second choice. They disagree on whether a third choice is acceptable and agree to meet again if turned down by the top two.

Such scenarios could also describe a religious body seeking a
new spiritual leader (pastor, rabbi, priest, imam), except that the
certain elders (by whatever name) would oversee the process, the

Making Better Choices. Charles E. Phelps and Guru Madhavan, Oxford University Press (2021). © Oxford
University Press. DOI: 10.1093/oso/9780190871147.003.0003

"job talk" would have the candidates conduct a weekly religious service, and the doctoral degree and publication record in metaphysical poetry would be replaced by proper theological credentials and experience. The same idea applies to private clubs seeking to choose new members to fill a few vacant slots created by previous members' moving from the region or by death. The selection criteria would differ some, but the process could be similar.

Consider another scenario: The incoming board chair or president believes that the organization lacks a clear focus and calls for a strategic planning process. The leader creates an ad hoc strategic planning committee, and the committee members are authorized to hire a consultant or facilitator to assist in their efforts. They talk to various professional colleagues, get bids from three firms, and hire a consultant. After due diligence and interviews with every board member (separately), the consultant convenes a two-day retreat at a plush golfing resort, where they create a standard SWOT chart to gauge their strengths, weaknesses, opportunities, and threats, and offer some ideas. The consultant then has the participants express their enthusiasm for each of the loosely described options by distributing 100 bright orange dots on the paper charts where each spitballed idea is posted on the wall. The consultant then takes this information, develops detailed strategies to fill out the top three of the ideas generated in the retreat (based on the dot votes), and presents them at the next board meeting. The board members rank them and adopt the highest-ranked alternative as their new strategic plan.

Here's yet another scenario: A federal government granting agency or a private foundation has issued a call for proposals for research projects to receive funding. The organization has committees of experts in the field of study assembled on a regular basis to review such proposals and rank them in priority to receive funding. Following their custom, they assign three of their members to carefully evaluate each proposal received. After each proposal has been presented and discussed, the panel members score the proposal using a best-to-worst 1-to-5 scale. Some of the members forget that 5 is worst and score inappropriately. The scores are averaged, and the average score creates the priority ranking they provide to the parent organization. Such a process could also easily describe an ad hoc panel of experts choosing the site for a new large-scale international research facility (such as the Large Hadron Collider or a new telescope array) or for hosting the Olympics, where the goal would be not to rank alternatives but to choose one winner.

And finally, consider this scenario. Some members of the homeowners' association have requested an expansion of the recreational facilities available to members and their guests to include, if possible, an indoor swimming pool (they live in a region with four real seasons, not Florida), squash and handball courts, and a year-round bar and meeting room area available for members to use for private functions. The directors of the association convene a special committee to consider the request. The committee members poll the residents for their specific interest in each of these functions and report back to the directors their summary of the general interest in the various options. They agree to proceed to get cost estimates for the swimming pool and bar or meeting area but not for the squash and handball courts (due to limited interest). Armed with this information, the directors present cost estimates to the full membership at their next annual meeting, where a vote is taken. They agree to hire an architectural firm to provide plans and cost estimates. A new committee is appointed to choose the architectural firm and supervise its work. The firm produces three different approaches (with different costs and degrees of opulence). The members then vote for their favorite. No single plan receives a majority. The board chair then proposes a runoff election between the top two. The second-ranked choice in the first vote receives the majority vote in the runoff. A member complains that the runoff vote is illegitimate, not being authorized in the association's bylaws. The objections are overruled, and the association proceeds to choose a contractor to carry out the work. A lawsuit ensues to block the construction.

Most decision scenarios tend to feature a common set of steps. Even though their order might differ, the steps generally include setting the goals of the process; designating a group to guide the process; specifying desirable attributes for options or candidates, and a method to identify and screen them; producing a shortlist; creating a set of finalists; and making the choice. What's often missing is the awareness and appreciation of the rules that underlie these processes or the eventual decisions.

At every step along the way, a set of rules—explicit or implicit, formal or informal—influences the decisions. In politics, they are constitutions. In organizations, they are bylaws. In other instances, they are codes of conduct. Elsewhere, they are compliance rules. Sometimes when they are not codified in writing, they are customs—"the way we've always done it." These rules can often predetermine the outcome of the process. Indeed, most people in most organizational settings have paid little to no attention to their rules until a major decision looms, at which point they turn to the bylaws to learn how the process should work. Having given little to no previous thought to the bylaws, people blindly follow them to guide the major decision they confront. After making that decision, they return to their usual activities until another major decision looms, at which point their attention again returns to the bylaws. But seldom do they ever look carefully at the bylaws themselves to determine if they serve the organization well.

Bylaws also specify the procedures through which individuals' preferences are combined into organizational decisions. They specify which members can vote, the type of ballot to be used in voting, and how those votes are assembled into a group choice. Members choose governing boards through such processes. The bylaws typically specify how meetings are to be conducted, including discussions and votes regarding specific actions the organization might take (he subject of a later chapter).

Bodies to whom decisions are delegated also have rules as to how they make decisions. These rules typically specify how many board members must be present to legitimize any decisions they make (a "quorum") and the process by which decisions are made (typically a majority of those present and voting). That simple statement guides how choices are framed for the board. If the decision requires a majority approval, then typically decisions must pose two and only two alternatives. Alas, many decisions do not neatly fit into this yes/no paradigm, so other methods must be used to convert complex decisions into yes/ no decisions if "a majority of those present and voting" is the prescribed decision rule.

Decision processes also contain some sort of filtering process to narrow down a large set of candidates into a manageable pool. This filtering has great value in many settings. If you have 47 candidates for one position, gathering detailed information about all of them can be an onerous task. Decision authorities (say, the board of directors) cannot be expected to evaluate large numbers of choices and choose among them. The ability to rank choices rapidly becomes impossible as the number of choices rises. Thus, organizations confronting such issues invariably delegate initial screening power to some smaller group to provide the filtering.

To see why this happens, think about the process of directly comparing three choices, A, B, and C. You compare A versus B, B versus C, and A versus C; with three candidates, three direct comparisons are needed. With four candidates, you compare A versus B, A versus C, A versus D, B versus C, B versus D, and finally, C versus D; thus, four candidates require six comparisons. Five candidates require 10 comparisons. Ten candidates require 45 comparisons, and 47 candidates require 1,128 comparisons. One hundred candidates require almost 5,000 comparisons. So careful evaluation of choices requires some sort of prescreening.

Prescreening processes are widely used but are vaguely defined at best. It's difficult to meaningfully catalog the range of such processes, because they are often secretive. In some organizations where we have asked to learn about the rules for such selection processes, we have been referred to legal counsel! Politely put, these rules are the "black box" of organizational decision-making.

One concrete suggestion regarding prescreening processes is to make them more transparent, or at least subject to subsequent review. In the next chapter, we discuss a variety of methods to grade or rank choices. These grades or scores could be made available for subsequent review and audit, processes that could help organizations determine their satisfaction with their own prescreening processes.[1]

---

1. There is even an experiment that organizations can readily conduct: Have two or more separate committees independently grade or score the entire set of candidates, and then see how well the conclusions match. An entire literature exists on the concept of inter-rater reliability.

Prescreening or not, rules are sometimes subject to the influence of an individual or a very small number of people within organizations. One publicly available—and typical—example is from the U.S. Chamber of Commerce. Article IV of its bylaws discusses how the nominating process works. The chair of the board designates a list of members for the nominating committee, *subject to the approval of the board.* Then the chair of the board designates the chair of the nominating committee. The board approves the committee membership. Think about this in light of the status quo bias and the power of framing. The chair is likely to have considerable say over the committee that nominates new board members, and total control over the chair of that committee. The committee chair then has ongoing control of the agenda, discussion, and votes of that committee. Of course, one of the key jobs of the board is to select its own chair for a term so the chair has considerable indirect and direct influence over the very people who choose the chair. The chair's powers are more extensive, as listed in Article V:

> The chair shall serve as the chief elected officer of the chamber of commerce and shall preside at all meetings of the membership, board of directors, and Executive Committee. The chair of the board shall, with advice and counsel of the president, assign the vice chair to divisional or departmental responsibility, subject to board approval. The chair of the board shall, with advice and counsel of the vice chairmen and the president, determine all committees, select all committee chairs, and assist in the selection of committee personnel, subject to approval of the board of directors.

Article VI, Section 1, discusses appointment authority:

The chair of the board, by and with the approval of the board of directors, shall appoint all committees and committee chairs. The chair of the board may appoint such ad hoc committees and their chair as deemed necessary to carry out the program of the chamber. Committee appointments shall be at the will and pleasure of the chair of the board and shall serve concurrently with the term of the appointing chair of the board, unless a different term is approved by the board of directors.

This obviously puts considerable power in the hands of a single person—the elected board chair—that can be reinforced through time. This may be good—indeed, it may be necessary for the functioning of the organization—as long as the concentration of power is a matter of comfort for the organization and the individuals who compose it.

In other instances, the concentration of decision rights is even denser. The U.S. president has enormous executive powers, held in check not by judicial process but only by the congressional process of impeachment by the U.S. House of Representatives and subsequent trial by the U.S. Senate. The Church of Rome has power concentrated in the pope in unusual ways, including the belief that when the pope issues certain documents (speaking *ex cathedra*), he cannot be in error.

Other organizations have much more diffuse power structures, originating with the traditional Athenian democracy, wherein every citizen (non-slave males over age 20) of Athens voted directly on matters of consequence—not through elected representatives. A quorum of 6,000 was established for important votes. A show of hands (majority rule) determined the outcome of each question. Later, to reduce the operational burden, Solon created "boule," a subset of the citizenry of 400 members selected by lot (soon expanded to 500). Further refinement of the concept established something closer to a representative government as we now recognize it.

Perhaps the closest to the Athenian democracy concept in current times—but without formal voting—is the congregation of the Society of Friends ("Quakers"). As they describe it:

> Once a month, the meeting (congregation) holds a "meeting for worship for business." Anyone who is part of the meeting may attend. Decisions are made without voting. Instead, the participants discuss the matter and listen deeply for a sense of spiritual unity. When the clerk recognizes that unity has been reached, it is called the "sense of the meeting." If those present agree with the clerk's expression of that sense, then the decision is recorded in the minutes.

Other sources say that this approval comes by members nodding their heads and saying "approve." Selection of the clerk of the meeting comes by approval of the members, with a nominating committee, the membership of which is also approved by the meeting. In modern society, seldom does the occasion arise where everybody can "come together" to discuss and vote on every issue. Instead, we choose decision-makers to decide on our behalf. We also approve the rules by which they operate—the bylaws.

Bylaws affect choices. Bylaws determine who has decision rights. Bylaws determine how collective choices are made, and by those who have decision rights, in ways that often go unnoticed. We will subsequently discuss how voting rules work, how well or poorly they can convey information from voters to decision-makers, and, indeed, even how they structure discussion and debates about choices made by organizations.

Every organization should carefully consider its bylaws and how the power is concentrated. They affect not only the leaders and employees of an organization but also everyone the organization serves or influences. Do the bylaws allow voters to express their views well, or do they stifle expression? Finally, do the bylaws restrict and control the nature of discussion and debate by artificially turning complex problems into a sequence of simple yes/no choices that are resolvable by simple majority votes? If so, what processes determine the transformation of complex problems into sequences of yes/no votes?

Most organizational bylaws were adopted at the time the organization was created, often by simply appropriating language from bylaws of similar organizations or from some template. State law to form a new corporation typically requires two documents: articles of incorporation and bylaws of operation. In some cases, as in homeowners' associations and charitable foundations, further laws specify many components of such bylaws.

The time to review an organization's governing documents comes *before* serious and complex decisions must be made, not when a major decision is upon it.

## Chapter 4

Choosing How to Choose

Majority wins in a democracy. Majority voting is easy to understand. Majority voting is easy to describe. Majority voting is easy to conduct. What could be wrong with this? The problem begins with the ballot itself. Any voting system has two major parts—the nature of the ballot (what can you say) and how that information is used (the algorithm to combine votes into a choice). A fundamental problem is that majority voting doesn't always determine a winner if three or more choices are on the ballot. Then the concept of "majority wins" becomes the one with the most votes wins or a runoff. Each can create amazing mischief. Scholars of social choice like to define criteria for each voting system and see how and where these criteria are violated. Passing certain criteria guarantees failures on others—some of these criteria, as we shall see, are mutually incompatible. Let's start with an example.

Suppose you and a group of 18 others are deciding on dinner options for an evening. You have four cuisine options: American

Making Better Choices. Charles E. Phelps and Guru Madhavan, Oxford University Press (2021). © Oxford University Press. DOI: 10.1093/oso/9780190871147.003.0004

(A), Brazilian (B), Chinese (C), and Delhi (to reflect Indian; D) restaurants.

You each vote on a ballot with a rank-order list, something like: rank #1 for Brazilian (B); rank #2 for Delhi (D); rank #3 for American (A); and rank #4 for Chinese (C). We'll describe these by saying B > A means that B is preferred to A. With B > D > A > C, the statement is that you prefer Brazilian to Delhi, which in turn you prefer to American, and in turn you prefer to Chinese. But we don't know by how much you like or dislike each of these choices since it's just a rank order.

If we had a numeric scale, you might have given the following values on a 0-to-100 scale for the four choices: 100, 99, 75, 15. This would tell us that you had no strong preference between the first two—the choices were nearly identical to you. American was somewhat further away, with a Chinese meal being your last preference, and not a happy one. Another group member might have the same ordering of preferences (B > D >A> C) but, on a numeric scale, might have had 95, 60, 60, 55.

While both of these voters prefer Brazilian first and Delhi second, the difference is almost irrelevant to the first voter but important for the second one. But rank-order lists cannot provide that information. Further, the second voter actually likes Delhi and American the same, but rank-order lists normally exclude "ties." So, this voter would have to flip a coin to choose which way to rank Delhi and American.

Comparing these two number scales emphasizes again the problems of adding up different people's scores. The first voter appears to use almost the entire range of 0 to 100 points. The second seems to compress the scores into a 50- to 100-point range, as if anything below 50 was unacceptable. Why do people use numbers differently? We can't be sure.

Voting rules (algorithms) combine the messages from each voter into a collective statement of desire. But economists limit the ways they think as legitimate to combine individual preferences. One approach for them is to limit the information to rank-order lists, called "ordinal" measures of utility.[1] Thus comes the dominant use of rank-order inputs in the general analysis of voting rules. Kenneth Arrow, a prominent economist, created what's called the "impossibility theorem" about voting, focusing entirely on rank-order "messages" from voters. Why did he focus on rank order?

1. Measure theory describes four main categories. Those with the lowest information content are "nominal"—the names of categories, like "animal, vegetable, mineral." They group things together, but with no implied order or sequence. These include modern things such as ZIP Codes and area codes, although these numbers (particularly ZIP Codes) imply certain geographic relationships. Next comes "ordinal" measures; just what they sound like, they indicate an order of sequence: 1 to N. But they don't say how far apart or close together they are; all they show is rank order. Then comes "interval" scales, where the numeric value has a specific meaning. Thermometers are interval scales. The "distance" between 22 degrees F and 25 degrees F is 3 degrees F, as is the distance between 35 degrees F and 38 degrees F. The distance between 22 degrees C and 25 degrees C is 3 degrees C, as is the distance between 31 degrees C and 34 degrees C. But 3 degrees C is not the same as 3 degrees F, so a distance of "3 degrees" doesn't mean the same. Nor does a level of "20 degrees" mean the same thing in both systems. As we all know, 20 degrees C is a lot more comfortable than 20 degrees F. Intervals have very different meanings. Finally, we come to "ratio" scales: Here, 20 is twice as large as 10, and 60 is 6 times as large as 10. That's not true with interval scales: 20 degrees F is *not* twice as warm as 10 degrees F. Ratio scales have a "natural zero" that's meaningful, not arbitrarily assigned.
  Both interval scales and ratio scales use "cardinal" numbers—numbers that say how many "things" there are. Ordinal numbers simply indicate sequence. In cardinal numbers, the distance between the numbers has meaning; not so with ordinal numbers. Within the class of cardinal numbers, ratio scales contain one additional bit of information: Zero is well defined. It means "none." Economists distrust using cardinal measures of utility because (like thermometers) it might involve comparing measures that in fact differ greatly one person to another.

Like most economists, he presumed that voters chose among candidates based on the "utility," a measure of happiness, that having each candidate in office would provide. Once you treat voting messages as measures of "utility," the die is already cast.

Because Arrow didn't trust cardinal measures of utility, he relied on rank-order lists, as woeful as they might be in conveying information. He then mathematically proved the following result: When you have three or more choices, for voting methods using rank-order ballots, *no* voting method can exist that fulfills the following four simple and desirable criteria:

1. Unanimity: *If every voter prefers the same candidate (say, candidate B), then the voting rule ensures that B wins the election.*
2. Universality: *The voting rule produces a winner no matter what the distribution of preferences among the voters.* Put differently, there does not exist some weird set of "tastes" that makes it impossible to determine a winner.
3. Rationality: *There is no rank reversal.* Rank reversal means that if one candidate enters or leaves the election, the final rankings of all other pairs remain unchanged. In our restaurant choice, if we learn that the Chinese restaurant is closed, the rankings between all other options in the final vote tally don't change. This is called "independence of irrelevant alternatives," and a lot of voting rules fail this rule.
4. Non-dictatorial: *There is no dictator whose vote determines the outcome.*

If the first three are satisfied, then the voting rule also requires that some dictator can always determine the election's outcome if need be. This can happen when the voting rule doesn't produce a winner, as with majority voting where there are three or more candidates. Or, turning it another way, if there is no dictator, either you have rank-reversal problems or the method doesn't always produce a winner. Even if every voter is wholly rational concerning their rank order, the voting process is not. If introduction or elimination of candidates can reverse the rank order of two other candidates under consideration, then the group voting becomes irrational.

Here are some of the many ways people have thought of to combine rank-order lists into a social choice.

The Marquis de Condorcet was an 18th-century mathematician, philosopher, and political scientist and a member of the French Academy of Science. His rule is simple and compelling: Compare each pair of candidates as if in a one-on-one election (directly from rank-order lists). If one candidate has a majority of the votes over all other candidates, that candidate wins. Condorcet voting is the expansion of majority rule to a situation with three or more choices.

Although both Condorcet and Arrow used rank-order lists as the basis for their analysis of voting, they had different ideas about what the rankings meant. While Arrow treated them as ordinal measures of utility, Condorcet treated them as the voters' perception of the candidates' merits. This sounds democratic. Indeed, who can challenge the logic that somebody who wins every pairwise vote should win the election? This criterion is one of the most prominent and widely admired ones to evaluate voting rules,[2] but the Condorcet rule doesn't always produce a winner.

2. Condorcet was not the first to develop this idea. Ramon Llul, a mathematician, theologian, and eventually martyr stoned to death in North Africa as he attempted to convert people to Christianity, described in 1299 a pairwise voting system for selecting the best candidate for religious leadership positions.

This turns out to matter more than one might imagine. Consider the children's game *Rock, Paper, Scissors*. That game has three rules that one can think of as voter preferences: Rock breaks scissors (R > S); scissors cut paper (S > P); paper covers rock (P > R). With these preferences, there is no winning choice. Any choice can be defeated by another. They are called "Condorcet cycles." It is easy to construct sets of voter preferences that create a cyclic voting pattern. Suppose there are k voters divided into three equal-sized groups with the preference patterns as A > B > C; B > C > A; C > A > B.

This set of voters' preferences produces a classic Condorcet cycle: Candidate A beats candidate B (the first and third sets of k voters), C beats A (the second and third sets of voters), and B beats C (the first and second sets of voters). This is, of course, the same problem as *Rock, Paper, Scissors*. When a Condorcet cycle exists, there is no meaningful winner.

To deal with this, Copeland proposed a method in the early 1950s to define a winner even when Condorcet's rule does not. He specified that for each candidate, the number of wins (vs. others) minus the number of losses be calculated for each candidate, with the largest value being the winner. This is akin to a round-robin sports tournament where every team plays every other team, and the one with the best record at the end of the tournament wins.

Unfortunately, Copeland's method also often fails to produce a clear winner, since two or more candidates often share the same winning total. The most simple and obvious case occurs with three candidate majority rule cycle, then each candidate has one loss. Sports leagues that use round-robin formats have extensive fallback methods to declare winners in the frequent event of ties, commonly focusing on such things as aggregate point differentials in head-to-head games or some variant. The fact that they often need to use such methods signals also that Copeland's method will commonly fail to resolve the issue. His method therefore has little meaningful value as a clear choice for a voting method. The puzzle remains what to do if no clear winner emerges.

Jean-Charles de Borda was a colleague of Condorcet in the French Academy of Sciences. A mathematician, physicist, and mariner, Borda fought as a naval officer against England in the U.S. War of Independence and had five French ships named in his honor. Borda proposed a rule in 1784[3] that was used until 1798, when Napoleon Bonaparte was elected to the academy and forced removal of the Borda process for reasons that remain mysterious.

Borda's rule has wide use today. This rule, or close variants, is used to rank U.S. collegiate sports teams (most notably football and basketball) and to choose "most valuable player" (or pitcher, rookie, etc.) in many professional sports leagues and the famous Heisman Trophy for the best collegiate football player. It has wide use in many other settings as well, and many people believe that it is synonymous with rank-order ballots. This is not true.

The Borda count is disarmingly simple. You assign points to candidates in accord with their rank on each ballot and then add up the points. For example, with six candidates, rank #1 gets 5 Borda points, rank #2 gets 4, and so forth, until finally rank #6 gets 0. To find the winner (and a rank ordering of the choices) you add up the Borda points from each ballot, the largest total being the winner.

The Borda points represent the number of candidates "bested" by each candidate on each voter's list. These methods are called sum-scoring methods, since they add up the scores associated with the ranks. Many variants on this approach exist, some adding bonus points for being ranked first, for example.

---

3. The concept developed by Borda in 1784 was previously described by Nicholas of Cusa (a.k.a. Cusanus) in 1433 CE. Nicholas was a philosopher, theologian, and astronomer.

Among its other faults, the Borda count (and any other sum-scoring algorithm) fails the Arrow rule of "rank reversal." It's easy to understand why. Two things happen when a new candidate enters the field. First, the total number of candidates increases, so the number of points assigned to each rank (except last place) go up by one. Second (depending on where the new candidate ranks on each ballot), some candidates get bumped down one rank. Those ranked above the new entrant (on each ballot) gain one Borda point, while those ranked below lose one rank. This can obviously change final Borda counts and hence rankings between candidates.

The same thing happens in reverse if a candidate leaves the field. Borda scores go down by one (in general) and some go up by one (if they had been ranked below the dropout on some ballots), thus offsetting the overall decline in Borda scores. So in this case, those ranked below the dropout generally benefit at the expense of those ranked above the dropout. An example that follows in Chapter 7 (the U.S. presidential election in 2000) shows vividly how this rank-reversal process can change voting outcomes.

Beyond the Condorcet rule and the Borda count, many other proposed ways to use rank-order lists exist to determine a winner. Here are a few of the most common (omitting some that are so complex that they require an advanced degree in math to understand how they work):[4]

---

4. For example, the Schulze method requires the use of graph theory, a branch of higher-order mathematics, and use of the "widest path" concept, best done using the Floyd–Warshall algorithm. Did you get that? The Kemeny–Young method counts all of the times different pairwise preferences exist in the ballots (e.g., A > B, A = B, and A < B) for all candidates. The method then counts how many ballots concur with or oppose each ranking. It is equivalent to minimizing Kendall's tau, a statistical measure. These counts determine the winning ranking, and the top name on that list is the overall winner. Got that also? The Kemeny–Young method is so complex that no computer algorithm yet devised can compute all winners in all situations. The problem is known as "N-P Hard," a way mathematicians describe really complex problems.

Black's rule. This rule tries to find a Condorcet winner. If none exists, the rule recommends the use of the Borda count. Many other combination voting rules also exist: The Condorcet rule is used first, and then if no Condorcet winner exists, some other method is used to find a winner.

Nanson rule. This rule proposes to rank all the candidates by Borda counts, drop all candidates with below-average Borda counts, and rescore the Borda counts. Keep doing this until you have a winner. Rank-order lists can be inferred from sequences of elimination. This is one of several methods that serially eliminate candidates.

Baldwin rule. This is identical to the Nanson rule, except you sequentially drop only the candidate with the lowest Borda count. Baldwin's method requires some sort of tiebreaker (e.g., a coin flip) if two or more candidates are tied with the lowest Borda count, or else you drop all who are tied with the lowest Borda count.

Instant runoff rule. This rule is used in a number of real elections, including to select the Australian House of Representatives and most Australian state legislatures; the legislative councils in India and Papua, New Guinea; and the president of Ireland. In the United States, it is used in Maine's House and Senate elections and became available experimentally in Utah for local elections beginning in 2019. The instant runoff rule is moderately complex. First, you seek a majority winner among candidates ranked #1. If one exists, that's your winner. But if there isn't a majority winner, you drop off the candidate with the fewest first-place ranks. Then all ballots are reranked without that candidate, again seeking a majority winner (using those ranked first) among the remaining choices. For those ballots who had ranked the dropout as #1, their #2 moves into their first rank, adding those to the #1 rankings of those candidates. This procedure is repeated as needed until a majority winner emerges. A number of variants have come about in recent years that allow more than a single vote transfer. In those more complex processes, the basic premise is that once a candidate with the majority of votes is elected, "excess" votes get transferred to second choices.

Coombs voting. This is similar to instant runoff voting except you drop off the candidate with the most last-place rankings (instead of using the fewest first-place rankings). The ballots are then retallied. These two methods have different effects. Instant runoff voting rewards polarizing candidates who have a core of strongly supportive voters but harms "compromise" candidates who are seldom voters' first choice but are often their second choice. Coombs voting removes highly polarizing candidates who have a strong core of enemies, but it preserves second-choice candidates in the pool.

Bucklin voting. This approach has still a different rule. It is not an elimination rule. Using rank-order lists as the ballot form, it asks, "Is there a majority winner with #1 ranks?" If so, that candidate wins, but if not, it then asks, "Is there a candidate who is at least the #2 choice of a majority?" If so, we have a winner. If not that, then at least the #3 choice determines the winner. Repeat as necessary.

Now we have eight ways to determine a winner from the same rank-order lists: Condorcet, Borda, Black, Nanson, Baldwin, instant runoff, Coombs, and Bucklin. These seemingly similar voting rules—in this case with all of them using the same ballots—can produce very different outcomes.

Back to choosing a restaurant. Where do you go for dinner? It all depends on which voting rule you use. You and your 18 friends all have turned in your rank-order ballots. We'll use a simple pattern of 19 voters' preferences found on an educational resource on-line about voting. There is nothing remarkable about this pattern of preferences, and many others can be found that would produce equally bizarre results. Here's the voting table:

	# Voters		
7	5	4	3
A	B	D	C
B	C	B	D
C	D	C	A
D	A	A	B

The number at the top of each column shows how many voters have a specific pattern of preferences, and the sequence down the column shows their ranking. Thus, there are seven voters (per the first column) with a preference ranking of A > B > C > D, five with preferences B > C > D > A, and so forth. We can loosely characterize these cuisines as follows:

A: Very polarizing, with seven votes at the top, many near the bottom.

B: Favorite of only five, but at least second choice among 16 of the 19 voters.

C: Favored by only three, otherwise middle of the pack (say, first or second compromise) but never last.

D: Spread widely; disliked by some, loved by some; similar to A, but not so extreme.

Now comes the dark magic of social choice theory. Let's see how our various voting rules turn these rank-order preferences into a collective preference about where to dine. **But before that, jot down what you think are the fairest method and the least fair method of these eight voting methods.** You may be surprised when you see the results.

CONDORCET RULE

B beats C (16 of the 19 voters, losing just those three in the last column) and also beats D (12–7, losing the seven voters in the last two columns). But A beats B (10–9, combining the first and last columns' voters). Alas, the Condorcet rule gives us no winner.

BORDA COUNT

Remember that a candidate gets $k - 1$ votes for first place (with k candidates). So with four candidates, first place is worth 3 points, second place gets 2, third place gets 1, and last place gets none. Here are the calculations:

A: $7 * 3 + 5 * 0 + 4 * 0 + 3 * 1 = 24$
B: $7 * 2 + 5 * 3 + 4 * 2 + 3 * 0 = 37$
C: $7 * 1 + 5 * 2 + 4 * 1 + 3 * 3 = 30$
D: $7 * 0 + 5 * 1 + 4 * 3 + 3 * 2 = 23$

Using the Borda count, the ranking is B > C > A > D. Everyone goes to the Brazilian restaurant.

BLACK'S RULE

With no Condorcet winner, Black's rule uses the Borda count, which chooses Brazilian.

NANSON'S RULE

Nanson's rule drops every choice with below-average Borda scores. The average Borda score is 28.5, so we drop A and D. Now we retabulate the voting table with A and D deleted:

# Voters			
7	5	4	3
B	B	B	C
C	C	C	B

It's obvious that B wins. The final-step Borda count is 16 for B, three for C. With just two candidates, the Borda counts equal the number of times each candidate is in first place.

The full rank ordering for the Nanson rule is the same here as the Borda count in this case: B > C > A > D. (It doesn't always work out this way.)

BALDWIN'S RULE

Baldwin's rule drops the single candidate with the lowest Borda count (instead of the very similar "below-average" Borda counts of Nanson's rule) and does the Borda count again. In the second round, after dropping D the distribution of preferences is:

# Voters			
7	5	4	3
A	B	B	C
B	C	C	A
C	A	A	B

The Borda counts (skipping the zeros) are:

$$A = 7 * 2 + 3 * 1 = 17$$
$$B = 7 * 1 + 9 * 2 = 25$$
$$C = 3 * 2 + 9 * 1 = 15$$

B still has the highest count, but we are not done, since we have three candidates still remaining. The final step drops C (the lowest Borda count with 15), and then we recalculate. The final voting table has this distribution:

# Voters			
7	5	4	3
A	B	B	A
B	A	A	B

The Borda counts are 10 for A and nine for B. (Remember that with only two candidates, the Borda count equals the number of first-place votes.)

In what seems like a trivial change in the voting rule between Baldwin and Nanson, now we have chosen American instead of Brazilian! American has climbed from third to first place. We have the same stated preferences and the same ballot system (rank-order lists using Borda counts to remove weaker candidates) but a wholly different outcome. Strangely, American wins despite having the second-lowest Borda count in the initial lineup, barely avoiding elimination in the first round.

INSTANT RUNOFF

Now we shift away from the Borda count as the way to drop "losers" and instead drop candidates using a different rule until we have a majority winner. In instant runoff, if there is no immediate majority winner, we drop the candidate with the fewest first-place votes. In this set of votes, there's no immediate winner, since nobody has a majority (A comes closest with seven but needs 10 to win). Since instant runoff drops the candidate with the fewest #1 rankings, C (only three first-place votes) is dropped and the new voting table looks like this:

# Voters			
7	5	4	3
A	B	D	D
B	D	B	A
D	A	A	B

Still nobody has a majority (10) of first-place votes. A has seven, D has seven, and B now holds the short straw (five first-place votes) and gets dropped. Now the voting table looks like this:

# Voters			
7	5	4	3
A	D	D	D
D	A	A	A

So with this rule, D wins the final vote, 12–7. We seem headed to Delhi food for some Indian spices. Strange? You betcha! The choice with the lowest Borda count (and the first to be dropped in the Coombs rule) actually wins the election. And the winner with the Borda count and Nanson's rule (Brazilian) was the second choice to get dropped.

COOMBS RULE

This rule works just like instant runoff, except that we drop the candidate with the most last-place votes (rather than the fewest first-place votes). In effect, this rule drops polarizing candidates with particularly heavy opposition to their candidacy. So, in the first round, we drop A, ranked last by nine voters, and the new voting table looks like this:

# Voters			
7	5	4	3
B	B	D	D
C	C	B	D
D	D	C	B

B immediately wins with 12 first-place votes, easily more than the 10 required for a majority.

If we were to sort out the complete priority list, D would be dropped next (12 last-place votes in this revised table), and the final sequence would be B > C > D > A. The group is eating Brazilian food with this rule. And Delhi food, which came in first in the apparently similar instant runoff, comes in third here.

This is another example of how a seemingly small change in the dropping rule changes the outcome. We saw it in comparing Nanson and Baldwin rules (using Borda counts) and it appears again here comparing instant runoff and Coombs. Again, these are not artifacts of a strangely designed set of voter preferences. It happens more . than one might imagine.

These four rules that drop candidates and vote again demonstrate vividly the "rank-reversal" issue identified in Arrow's impossibility theorem. As the different rules eliminate different candidates sequentially, the rank reversal issue looms large. If voting systems using rank order didn't have this rank-reversal problem, all of them would yield the same winner.

BUCKLIN VOTING

As shown in the original voting table, no simple-majority winner exists among votes ranked #1. How about first or second? B is ranked first or second on 16 of the 19 ballots, a clear winner with the Bucklin rule. If one were to do a full rank order of all four options, then we would set B aside and recount for second place. The voting table to determine the second place looks like this:

# Voters			
7	5	4	3
A	C	D	C
C	D	C	D
D	A	A	A

No immediate winner exists, but C is at least in second place in all 19 ballots (with B already chosen as #1). So Chinese is the second choice here. Choosing between A and D for third place, D wins 12–7 for third place, and A is fourth. The final sequence is B > C > D > A.

PLURALITY VOTE

Now, in addition to our rank-order ballots, let's go back to vote-for-one, assuming that the #1-ranked choice on each ballot is the vote-for-one candidate. If we use a plurality rule, A wins with the final ranking of A > B > D > C and seven, five, four, and three votes. Again, here's the voting table, and we basically just use the top row of it (shown in boldface), demonstrating how much potential information this voting system discards (un-bolded):

# Voters			
7	5	4	3
A	**B**	**D**	**C**
B	C	B	D
C	D	C	A
D	A	A	B

RUNOFF VOTE

In a runoff, if there is no majority winner in the first round, you have a second vote between the top two vote-getters. Here, the top two vote-getters are A and B. In a runoff between those two, A gets 10 votes and B gets nine, so this time, we get the same result as the straight plurality vote. If there were a second runoff (like the consolation game in a basketball tournament), it would pit C against D, and C would win 15–4. The final ranking would be A > B > C > D.

But what if the three voters in the last column are so un-happy with their situation (forcing them to choose between their two last-place preferences) that they don't vote in the runoff? Instead, they decide to leave the large dinner party and eat Chinese, leaving the other 16 to sort it out themselves. Now Brazilian wins 9–7.

A larger lesson resides here: If the set of choices affects voting participation, a runoff election might have unpredictable effects. In this example, a runoff will elect either A or B, depending on how the fans of Chinese react to the choice. If they were completely cyn-ical, they might even blackmail the fans of Brazilian to pay for their meals in order to get them to vote to overturn the plurality favorite: American.

Yet another system known as "contingent voting" asks voters for a rank-order list, and then immediately calculates the runoff winner the same as the two-election runoff described here. This saves the costs and time of the second vote and eliminates the possibility for some voters to drop out in the second election (or for others to vote only in the runoff). A variant of this system is used to elect the Sri Lankan president, wherein the top three choices are used in the runoff.

We can summarize these 11 outcomes, all of which use rank order:

Condorcet	No winner
Borda	B > C > A > D
Black	B > C > D > A
Nanson	B > C > A > D
Baldwin	A > B > C > D
Instant runoff	D > A > B > C
Coombs	B > C > D > A
Bucklin	B > C > D > A
Plurality	A > B > C > D
Runoff, contingent vote	A > B > C > D
Runoff with defection	B > A > C > D

The ranking of B > C > D > A seems to emerge more than any other—but in only three of the 11 outcomes. Rankings with B > C at the beginning occur most often (five of the 11 methods). B is ranked first on six of the 11 different voting rules. American wins with Baldwin or vote-for-one with both a plurality rule and a runoff (unless some voters avoid the runoff). If we use instant runoff, Delhi wins.

None of these methods elect C. But here's a voting rule that lets C win: Several real-world votes use a Borda count variant with extra "bonus points" for being in first place (most notably, votes for Most Valuable Player in Major League Baseball). One can make C a winner by creating a rule with 3 or more penalty points for being in last place. If one were to redo the Borda counts with −3 for last place (instead of zero), then C is the winner. It's easy to intuit how to do this, since C is never in last place on anybody's list. So penalizing "last place" must favor C. The only question is how big the penalty has to be. Using 3 penalty points for a last-place ranking does the trick with this particular pattern of voters' preferences.

Before we went through this example, we asked you to jot down which method seemed most fair, and which seemed least fair. How do you feel about these methods now?

In recent decades, the social choice analyses have moved beyond the sole use of vote-for-one and rank-ordered approaches. We next discuss several of these.

MAJORITY JUDGMENT

Developed by scholars Michel Balinski and Rida Laraki, majority judgment offers a novel way of "voting" and provides an escape path from Arrow's impossibility theorem. The basic idea of majority judgment is reasonably simple and familiar to most of us: Voters grade the candidates, just like teachers grade students. A central feature is the use of a "common language" of grades among the voting population—at least some general agreement about the meaning of the various grades, as with the following example:

A	Excellent
B	Very good
C	Good
D	Fair
E	Poor
F	Unacceptable

Majority judgment differs from rank-order and other methods, because the statement is not how much "utility" the voter gets if their candidate is elected, but rather each voter's evaluation of each candidate's fit for the job. (Condorcet's original method also put the voters' evaluations of "the merits of the candidates" into rank order.) Majority judgment eliminates rank reversal, since each candidate's grade by each voter is independent from that of every other candidate. Voters don't need to use all the grades, and they can give the same grade multiple times if they wish. Thus, voters can readily express both ties and gaps in evaluation, neither of which is present in rank-order lists.

What's a bit more complicated is the description of how the grades are combined to determine the winner. Unlike classes we took in high school or college with one teacher assigning us grades, we have many "teachers" grading many "students." So, we need to know how to combine their grades to get a collective choice. The majority judgment method seeks the grade that can receive a majority vote—a simple idea.

How to do this? The basic idea is to line up the grades for each candidate, and find the middlemost. You compare these median grades for each candidate. If there are no ties for median grades, then the candidate with the highest median grade wins. If two or more have the same median grade, then several rules come into play. They invoke comparison with a "scales of justice" model where all of the median grades stack up on the pivot point. All grades above the median grade sit in one scale pan, and all below the median grade in the other. The one with greater weight on the positive side wins. Note that in these rules, *how far* above or below the median grade doesn't matter—just whether they are above or below the median grade. This logic basically asks how much more "weight" there is in the two pans on the different sides of the scales of justice.

The median grade has one key feature: It's the only grade that can receive a majority vote agreeing that it is the proper grade. Majority judgment, in effect, asks people to vote on the grade each candidate should receive, with a "majority rule" deciding each grade, and hence the election outcome. "Judgment" comes from the grading. "Majority" comes from the implicit majority-voting process to select the final grade for each candidate. If you have an even number of voters, and the middlemost pair has two different grades, you use the lower of the two grades, because that lower grade is needed to gain a majority that the grade should be at least the chosen grade. This is the proper approach since higher grades defeat lower grades.

To see how this works, let's set up a table of hypothetical grades from our 19 dinner-companion voters that are similar to the ranking in our earlier "voter table." There are many possible sets of grades that are similar to the rank-order lists shown previously, so this is just an example to show how majority judgment works. The median grade is shown in bold font with nine grades on each side of the median. Here we show the letter grades A, B, . . . F in italics to distinguish them from the letters A, B, C, and D that we used earlier as abbreviations for the restaurant choices.

American	*A A A B B B B C C* **C** *C C C C D D E E F*
Brazilian	*A A A A A A A B B* **B** *B C C C C C D D E*
Chinese	*A A B B B C C C C* **C** *C C C D D D D E F*
Delhi	*A A A B B B B B C* **C** *C C C C C E E E F*

Arrayed from left to right, the string of letters shows the grades assigned by our 19 hypothetical dining companions. With this array of grades, Brazilian wins, having a median grade of B, while all other choices had a median grade of C. We could also describe these with a table like this:

Summary of grades for four dining choices

	A	B	C	D	E	F	Total	Median
American	3	4	7	2	2	1	19	C
Brazilian	7	4	5	2	1	0	19	B
Chinese	2	3	8	4	1	1	19	C
Delhi	3	5	7	0	3	1	19	C

Now, it's simple to find the median—add up along a row until you reach the necessary total (10 votes to get a majority with 19 voters). This gives the median grades shown in the last column. It doesn't matter whether you add from the left (A) or the right (F)— the median is the same either way. If the total vote count is even, then the median involves a pair of grades, in which case the lower of the two grades is used.

Resolving ties among choices with the same grade is modestly more complicated but adheres to the same basic idea, as the previously mentioned "scales of justice" analogy suggests (and Appendix A provides more details). In our example (excluding Brazilian, which has already won with a median grade of B), Delhi will come in second (with eight votes above the median grade of C), then American (seven votes above the median C), and the final ranking here is Brazilian > Delhi > American > Chinese.

The tie-breaking process becomes less tedious and the winner more obvious if the grading system allows graders to use "plus" and "minus" modifiers. If there are six grades from A through F, there are six possible median grades. With "plus" and "minus" modifiers on all grades (A⁺ to F⁻), there are 18 possible median grades, so the likelihood of a tie will drop precipitously, thus reducing the need to invoke the more complicated tie-breaking rules.

Further, among its other virtues, this median-grade approach is very robust against strategic voting, since a grade of D, E, or F has the same effect on the median compared with C. If you let "F" mean more than "D," you invite strategic voting. Chapter 6 discusses this issue in further detail.

Finally, looking back at the Bucklin method (which uses a voting majority approach), we can see that Bucklin voting is somewhat similar to majority judgment, except that the ranking involved in Bucklin voting does not allow ties in evaluations or gaps between the candidates' scores. If rank-order lists for Bucklin voting allowed ties and gaps, it would closely resemble majority judgment but would still lack the meaning of the categorical grades that the latter provides. Indeed, the developers of majority judgment recommend that if you are "stuck" solely with rank-order lists, the best choice is to use Borda majority-judgment median-based rules to determine the winner.

APPROVAL VOTING

The process is simple to describe and simple to use. Voters receive a ballot listing all of the candidates, and they simply put a checkmark in the box beside each of them that they "approve." No effort is made to define what "approve" means. Winners and rank order are determined by the total number of approvals.

Approval voting has both merits and concerns. It is immune from rank reversal, since the voting method applies a label to each choice independent of the others. In that sense, it's like majority judgment with only two grades. (It differs in the way the scores are combined.) Viewed this way, it is clearly inferior to traditional majority judgment, which typically uses about six grades, allowing much greater expressivity to voters.

Because approval voting sums the scores for each candidate, it falls into the category of cardinal voting systems, where numeric scores are added up. And just like any cardinal scoring system, there remains an important ambiguity—what does "approve" mean? Since the proponents of approval voting specifically do not define what the word means, we are left with the ambiguity of any cardinal system and the impropriety of adding up the numbers across ballots.

RANGE VOTING

This method is also simple to describe and use but comes with defects. Voters assign a score of 0 to 99 (or an agreed-upon range) to each candidate—a numeric grade akin to majority judgment's letter grades, but many more of them. But then scoring computes the average. Therein stems a problem: These are cardinal scores, so adding them up makes sense if and only if each voter has the same scale in mind. For example, does "100" mean the best candidate that the voter can envision, the best that the voter has ever encountered in person, or the best among the existing field? Similar questions appear about the meaning of "0."

Scoring provides obvious options for strategic voting, and the effect is stronger when the range of points is wider. The same effect arises in rank-order voting when the number of candidates rises. This is obvious, for example, when considering the Borda count, where the number of points rises for each rank except last place, which gets 0. So the penalty for being ranked last rises as the number of candidates rises. A similar issue occurs when the number of points allowed for scoring rises from, say, 20 to 50 or 50 to 100.

In one sense, approval voting is just scoring with only two scores allowed—0 and 1—instead of the wider ranges usually associated with scoring (0 to 99 or 0 to 20). Both methods add up the "points" and neither has a clear definition of what the numbers mean.

STAR VOTING

Amazon asks for feedback on sellers using five stars, as does Yelp for restaurants, home and automobile services, and many others. Even the Transportation Security Administration (TSA) asks for your rating of your experience with a "smiley face" gadget sitting inside the security area, just after you finish your airport security check. Uber carries it one step further, since passengers rate drivers *and* drivers rate passengers using a five-star system. These ratings have become a common mode of expression in our modern commercial world. In fact, it's just a limited-vocabulary scoring system, with scores of 1, 2, 3, 4, and 5. If you want to think about your experiences with scoring, think of star voting as a reasonable example of the system.

CUMULATIVE VOTING

We can trace the use of cumulative voting at least back to the 1850s in Europe. With this model, every voter gets a specific number of votes, which they can distribute among the available candidates. In most corporate board-of-director votes, if there are k slots open, each voter gets k votes. In the business use of cumulative voting (what consultants often call "dotmocracy"), facilitators commonly give each voter a pad with 100 brightly colored sticky dots, which they can use (wandering around a room with descriptions of various choices hanging on the wall) and put as many or as few dots as they wish on each option. At the end, you count up the dots to rank the choices.

Two themes emerge from the voting methods described. The first theme pertains to the older voting methods—vote-for-one (probably developed in the Athenian democracy some 2,500 years ago), rank order (going back to the era of Ramon Llull in 1299 CE), and cumulative voting. These ballots require that the voter compare all candidates while marking the ballot, either to pick the best of all or to rank them in sequence—call them simultaneous ballots. All of these approaches suffer from the risk of rank reversal.

While the newer ballots—approval voting, scoring, and majority judgment—do not share this problem, they create a different concern. The newer ballots require adding up the points assigned on each ballot or taking their medians. Doing this makes no sense unless the voters share a common vocabulary about the meaning of the points or grades they assign.

To see this, suppose you asked people to report the average temperature in the city where they live. People in Barcelona, Spain, might report monthly averages ranging from 12 to 30 degrees. People in Anchorage, Alaska, might report monthly averages between 17 and 60 degrees. If we don't know which temperature scale each city uses, the erroneous conclusion is easy to draw—that Anchorage is warmer than Barcelona!

Any person familiar with the different temperature scales of Herr Fahrenheit (the German) and Herr Celsius (the Swede) would immediately understand the problem, but with many voting methods, we have no way to know whether people are using the same thermometer. Plus, many voting systems automatically presume a common (even same) scale of measurement. This can lead to incorrect inferences.

Among the newer ballots, only majority judgment offers a common vocabulary. Approval voting seldom defines what "approve" actually means. Scoring can have a range of meanings. Cumulative voting does create a common vocabulary (with the constraint of allowable points), but it runs the risk of rank reversal since that very process requires comparing across candidates.

Looking back, this chapter may remind us of a now-quaint circumstance. A young school student goes to the town library to ask for books about penguins to prepare for a writing assignment. The librarian returns with a huge cart overflowing with books. The student looks at the collection and tells the librarian, "Thank you, but that's really more than I wanted to know about penguins."

This may be how you feel about voting methods at this point. But our hope was to enable an appreciation of how the different voting methods work and how one might choose a method that serves the best needs of a group. This takes us into the realm of "human factors"—the engineering concept of how useful, usable, and effective a voting method is, and how well it allows voters to express themselves.

## Chapter 5

Insincerely Yours

Many ways exist to alter the outcome of elections within the customs of standard voting rules. *Every* known voting rule can somehow be manipulated if people vote strategically (insincerely) to improve the chances of their favorite choice to win the election. Manipulation also can happen in the way we frame and sequence the available options. Results could be distorted by the kinds of moderators we choose for those discussions. These are prime matters of strategic voting. In strategic voting, some people actually don't report their true preferences. Rather than "voting sincerely" they vote strategically to improve the chances that their most-favored candidate will win.

Consider this famous example from the Cold War era. The Olympics created intense focus on the total number of medals won by various nations. The medal count became a symbol of national strength. These "proxy wars" alongside the Cold War activities were fought vigorously using many methods. Some countries used performance-enhancing drugs, most notably East Germany in an earlier era and Russia in the 21st century; the latter was even banned from competing in the Olympics for four years because of widespread state-organized doping.

Making Better Choices. Charles E. Phelps and Guru Madhavan, Oxford University Press (2021). © Oxford University Press. DOI: 10.1093/oso/9780190871147.003.0005

Races can be measured with timers; jumps and thrown objects can be measured by their distances. All of these are nearly immune to tampering—they are wholly objective measures. But some sports have no such standard, most notably diving, gymnastics, and ice skating and ice dancing. There, beauty is in the eye of the beholder. Obviously, this also applies to other contests, including music, theater, art, and similar endeavors for which no objective measure of value exists. Of course, ratings of foods and beverages (most notably wines) have the same issues.

The then Soviet Union coopted judges from East Germany, who notoriously over-scored Soviet performers and under-scored U.S. performers. Of course, the Soviet judges did the same thing for the East German skaters. The United States responded in turn by asking Canadian judges to do the reverse for U.S. and Soviet competitors, and similarly the U.S. judges favored Canadians. As a result, one could see a set of scores for a U.S. performer (on a scale of 10 as perfect) that ran something like 9.8, 9.6, 9.6, 9.4, 6.2, the latter being the East German judge. If the scoring method simply averages those numbers, the effect of the 6.2 can demolish that skater's chance of victory. Note also that by allowing decimal points, these skating scores have an equivalent range from 0 to 100, exacerbating the effects of strategic voting. The only limitation on the scoring came from mandatory point deductions of fixed amounts (say, 0.1 points) for specific errors (say, completing only a double jump instead of the programmed triple jump).

This practice became blatantly obvious in the 1998 Winter Olympics in Alberta, Canada. To avoid this, the International Skating Union adopted some simple changes. They first expanded the judge pool in size to 12. They then randomly discarded three scores. They also dropped the high and low scores. Then in 2008, as a cost-saving measure, they reduced the judge slate to nine and simply discarded the high and low scores.

All strategic voting requires some sense of what other voters intend to do. In that sense, political polls, for example, exacerbate the problem of strategic voting. In the United States, trying to ban political polls would violate constitutional protections of free speech. So we should operate under the presumption that people know "something" about others' intentions, perhaps imperfectly. This would include voter turnout forecasts as relevant information. With that caveat in mind, here are some more examples, returning to the "where to go to dinner" situation in Chapter 4. The preference profile looked like:

# Voters			
7	5	4	3
A	B	D	C
B	C	B	D
C	D	C	A
D	A	A	B

Suppose the prospective diners have agreed to use a vote-for-one ballot, with a runoff if nothing receives a majority. If everybody votes sincerely, we'll have no majority, and a runoff will pit American (seven votes) versus Brazilian (five votes). With sincere voting, A would win that runoff as well by a 10–9 score.

However, suppose the three voters most favoring Chinese food realize the lack of support for their preferred choice. They might simply refuse to vote and go off on their own for a Chinese dinner.

However, if they all agree to vote *en bloc* for D (their second choice) instead of C (their true first choice), then D will tie A for the runoff (seven votes each). In that runoff, D gets 12 votes and A gets seven. So, those three voters in column 4 can actually get their second-best choice (D) by voting strategically rather than sincerely in the first round. Indeed, only two of them have to change their votes, since D would then have six votes and be a finalist, at which point sincere voting would carry the day for them.

The obvious and frequent strategic voting in the United States happens when people can vote for only one candidate, and a plurality vote wins. They would prefer candidate C, but polling information tells them that C cannot win—either A or B will win. If the vote between A and B seems decisive, they may vote for C anyway, to express their sincere opinions, but if the race between A and B seems close, many who truly prefer C will vote for their second-best choice (either A or B) rather than "wasting" their vote on C. Their behavior hinges on expectations of other voters' behavior. Many other strategic voting opportunities hinge on rank reversal. In simple terms, if you can get rid of your most dangerous opponent in an earlier voting round, you can improve your outcome.

In voting systems with numeric scores that are added up (whether they are scores or numbers assigned to rank order), the wider the range of scores, the more power is given to strategic voting. Thus, for example, strategic voting becomes more of an issue in Borda-count (and related) elections when the number of candidates grows. A simple strategic vote gives your most favored candidate a rank of 1 and your most dangerous ("closest") opponent a rank of k. Your most preferred choice gets $(k - 1)$ Borda points, and your last-ranked choice gets 0. Thus, the bigger the field (larger k), the more impact this can have on the election. This is magnified even more in something like range voting, where scores typically run from 0 to 99. Any score-summing or point-summing method has this risk, and the bigger the numeric range, the greater the invitation to vote strategically.

Range voting can be easily manipulated by obvious strategic voting so that even in a simple two-candidate, three-voter election, the majority favorite loses a range-voting election. Consider this voting profile:

	Candidates	
Voter	A	B
1	100	0
2	50	100
3	50	90
Total	200	190

Here, voters 2 and 3 vote sincerely, preferring B to A, but give honest scores to A. Voter 1, however, strategically votes 100 for A and 0 for B, taking advantage of the wide range of scores, and gives A the election with range-voting rules by the total point count of 200 over 190. In general, point-summing methods are most susceptible to this, and the wider the point range allowed, the greater the risk.

Many organizations work hard to avoid strategic voting. In wine evaluations, the tasting is almost invariably done "blind" so the tasters have no idea which wine they are rating. While they may believe they know, in real wine tasting, even the top experts regularly fail in guessing the major grape and the country of origin. So blind taste-testing works well to remove strategic voting. This provides perhaps an extreme example of how strategic voting depends on knowing what other voters will do. In blind tasting, it is virtually impossible to forecast how other voters will score each wine, since they are all blinded. The information about other voters' intent is removed.

Strategic voting techniques are dizzyingly large. How can they be minimized? Here are some pathways.

First, be vigilant about conflicts of interest. This is a key topic in many organizational bylaws. Directors and trustees of corporations generally recuse themselves (or at least do not vote) during discussions of matters where potential conflicts exist. If you remove the *reason* for strategic voting, you reduce the risk that it will occur. Conflicts of interest are not behavior; they are situations. A person is said to be in a conflicted position when loyalties are due (legally or otherwise) to multiple parties of a transaction. Directors of corporations have a *fiduciary* obligation to their organization—the highest legal standard of trust and devotion of duty. Thus, fiduciaries should recuse themselves when they have competing interests. Almost all scientific journals now require disclosure of potential conflicts of interest for manuscripts submitted for possible publication. Conflict-of-interest disclosures do not prohibit publication of the research but instead alert readers to the situation.

In matters of voting, organizations can take an even firmer stand and require that conflicted individuals recuse themselves from voting. In stark contrast, international skating judges are still not forbidden from judging performers from their own nation. Some contests allow complete anonymity in judging, as noted earlier with wine tasting. This obviously cannot be done for many things, most obviously sporting events such as skating and gymnastics, art, and refereeing of other sporting events such as football, cricket, soccer, basketball, baseball, ice hockey, and the like.

A second way to lessen the effects of strategic voting is to trim the extremes before taking the averages. Our earlier discussion of how the International Skating Union has grappled with strategic voting touched upon the idea of trimmed means as a safety net against strategic voting.

Trimming of means is also used by central banks to provide the London Inter-Bank Offer Rate (LIBOR). Beginning with the interest rates charged by the top 18 banks in London, LIBOR removes the top and bottom four rates and publishes the average of the remaining 10. Many financial transactions around the world peg interest rates to the LIBOR daily quote. LIBOR trims the means to ensure that outliers do not influence their reported values. Another approach drops the extreme quartiles (the top and bottom 25 percent of scores) and then takes the average, called the "inter-quartile mean."

The median of a distribution trims away all observations except the middle ones. In completely symmetric distributions of data (like the classic "bell curve" of the normal distribution), the mean and the median are, in concept, the same value, but real data almost never have perfectly symmetric distributions. This feature of medians—serving as bulletproof protection against outliers—provides the same logic for thinking about rules that both avoid the effects of and deter the use of strategic voting.

The majority-judgment approach developed by Michel Balinski and Rida Laraki uses the concept of "middlemost" as the focal point. The middlemost in a distribution of grades wins the election in their model. Think of two judges. Judge 1 prefers A over B and Judge 2 has the reverse ranking. Then Judge 1 can either increase the ranking of A or lower the ranking of B, but cannot do both, and vice versa. Judge 2 can either raise the rank of B in parallel or lower the rank of A to combat the strategic voting of Judge 1. Thus, they tend to neutralize each other. In majority judgment, this is called a "partially strategy-proof" rule. This gives the greatest known protection against strategic voting.

Voting rules are not the only source of concern about strategic voting and manipulation. Indeed, some of the most important issues related to strategy in voting come when everything is boiled down to simple yes/no votes where majority vote reigns supreme. How can you get strategic manipulation when the vote is a simple yes/no? The next section dives into these issues.

In order to bypass the common problems associated with three or more choices, many organizations attempt to turn every complex vote into a sequence of binary votes—two choices, where one wins and one loses. At each step along the way in these processes, a simple majority vote controls the outcome. But these processes are also subject to manipulation. Individuals and committees can control the outcome of a voting process by the sequence of voting for the final choice. This is agenda control.

Legislative bodies work almost entirely through committees that represent the first step toward final approval. In the U.S. Congress, the key agenda-control steps take place in numerous subject-matter committees so that the final proposed legislation emerging from the committee often proceeds unscathed through final votes. Sometimes the proper agenda management requires an incomplete bill to emerge from committee (in order to get sufficient votes at the committee level) and then subsequent amendment in the final vote in the full House or Senate—where the distribution of preferences may differ from that in the committee.

Private organizations—social clubs, religious organizations, faculty groups, scientific academies, professional societies, businesses, and more—commonly use a method that resembles a parliamentary procedure. The most famous and widely used of these is *Robert's Rules of Order*. Henry Robert (1837–1923), a West Point graduate, was a major in the U.S. Army when he published the first edition of his rules. He went on to become the U.S. Army's chief engineer. He became motivated to do this when asked to lead a church meeting, for which he did not feel he was well prepared. His rules loosely followed those used in the U.S. House of Representatives at the time, so it is no coincidence that *Robert's Rules* and standard parliamentary procedure have strong similarity. Parliamentarian associations report that up to 95 percent of U.S. organizations use *Robert's Rules*.

These rules control the sequence of sequential votes to amend and alter proposals put forward for a vote. They apply to yes/no voting situations. Their primary purpose is to convert complex problems into a sequence of yes/no votes. In achieving this goal, a group can create a situation where the procedure (not the substance of the problem) dominates the discussion.

Consider an imaginary example of such a discussion, drawn from the fictional minutes of the fictional Shady Grove Country Club golf course revision debate.

PRESIDENT: *Today we have a report from the Course Revision Committee.*

COURSE REVISION COMMITTEE CHAIR: *Thank you. We have two proposed plans from our consultants, one to make the course slightly easier, one to make it slightly harder to play. We call these Plan Easier and Plan Harder. The third option is to make no alterations on our course.*

[Presentation of recommendations; discussion ensues.]

COURSE REVISION COMMITTEE CHAIR: *The Committee recommends that we accept Plan Harder and revise every hole accordingly, with the exception of Hole 7, which is the iconic symbol of our club, which would remain the same. I so move.*

[ANGRY VOICE FROM THE FLOOR]: *That's easy for you to say, Lynn! Wait until you're in your 60s and you'll want Plan Easier! You and that bunch of semi-pro golfers on the Committee didn't think about the broader membership!*

PRESIDENT: *Easy, now, Morgan. We'll have plenty of time to discuss this.*

[VOICE FROM THE FLOOR]: *I second the motion.*

[Further discussion ensues.]

[VOICE FROM THE FLOOR]: *I move to amend the motion to exclude Hole 13 as well as Hole 7 from change. It is just as important to our history as Hole 7.*

[ANOTHER VOICE FROM THE FLOOR]: *I second the amendment.*

[NEW VOICE FROM THE FLOOR]: *I move to amend the motion to say that we will use the Easier plan, not the Harder one.*

PRESIDENT: *That's out of order.*

[SAME VOICE FROM THE FLOOR]: *Why?*

PRESIDENT: *We have to vote on the amendment in front of us first. The best way to do what you want is to defeat the motion on the floor, and then you propose that as a new motion. But first we have to deal with the amendment.*

[SAME VOICE FROM THE FLOOR]: *OK. I think I get it.*

[ANOTHER VOICE FROM THE FLOOR]: *I like Hole 7 as it is. But I hate 13 as it is. How do I vote here?*

PRESIDENT: *You vote for the amendment and then against the motions. No, wait . . . you vote against the amendment and then for the motion.*

[SAME VOICE FROM THE FLOOR, FROWNING SLIGHTLY]: *OK. I think I get it.*

[ANOTHER VOICE FROM THE FLOOR]: *I want to keep Hole 13 as it is, but change Hole 7 according to the proposal. How do we do that?*

PRESIDENT: *I think we'd have to reject the amendment, then propose another amendment that would change Hole 7 following the Harder plan, but remove Hole 13 from being changed.*

[VOICE FROM THE FLOOR]: *I so move.*

PRESIDENT: *You can't do that yet. We have to vote on the current amendment first, and if it's defeated, then you can make your motion.*

[VOICE FROM THE FLOOR]: *What if it passes?*

PRESIDENT: *Then you can propose another amendment to keep the current Hole 13 but use Hole 7 from the Harder plan.*

[SAME VOICE FROM THE FLOOR]: *OK. Just don't forget me.*

PRESIDENT: *I think this would have been simpler if the original motion had just said to adopt Plan Harder, and then work from that to amend it. Would the Committee accept a friendly amendment to do that?*

PARLIAMENTARIAN: *We can't do that. There's another amendment on the floor that conflicts with that.*

PRESIDENT: *OK. Let's take a vote on the amendment. The vote is to add Hole 13 to the list along with Hole 7 that is not modified by Plan Harder. All in favor?* [Show of hands.] *All opposed?* [Larger show of hands:] *I'd better count that. All in favor, again?* [Show of hands again; specific count is taken.] *All opposed again?* [Specific count is taken.] *The motion to amend fails.*

[VOICE FROM THE FLOOR]: *I don't like Plan Harder. I call the question.*

[ANOTHER VOICE FROM THE FLOOR]: *Wait a minute! I didn't get my chance to make the amendment that I was promised!*

PARLIAMENTARIAN: *The question has been called. Is there a second?*

[VIGOROUS VOICE FROM THE FLOOR]: *I second it!*

PARLIAMENTARIAN: *We have to vote on that motion now. It is a privileged motion according to* Robert's Rules of Order.

PRESIDENT: *OK, sorry, Tracy, but we have to vote on that now. We have to decide whether or not we end debate on the motion.*

TRACY [MUTTERING TO SELF]: *What the #*^%&! is going on here?*

PRESIDENT: *OK, a show of hands to end debate on the motion. All in favor?* [It passes widely.]

PRESIDENT: *OK, now we have to vote on the original motion.*

[CONFUSED VOICE FROM THE FLOOR]: *Can you please restate the motion we're voting on?*

PRESIDENT: *We're voting on whether to accept Plan Harder or not.*

[SEVERAL VOICES FROM FLOOR, TOGETHER]: **NO!**

PRESIDENT: *Oh, I forgot, it's to accept Plan Harder but keep Hole 7 as it is now. All in favor?*

[Heavy vote against.]

PRESIDENT: *The motion fails.*

[VOICE FROM THE FLOOR]: *I move that we accept Plan Easier.*

[DIFFERENT VOICE FROM THE FLOOR]: *What about Hole 7?*

[THIRD PERSON FROM THE FLOOR]: *And Hole 13?*

[Debate continues.
They adjourn in frustration 3 hours later with no decision.]

Now instead consider how this would work in a parallel universe where the Shady Grove Country Club had adopted ways to vote on many choices also in parallel. In this setting, the discussion focuses on the substance and people can follow the discussion clearly and are not frustrated by the process. In such a discussion, the various issues would lead to consideration of four main alternatives with standard variants: (1) accept Plan Harder; (2) accept Plan Easier; (3) do half the golf course "harder" and half "easier"; and (4) leave it alone. This leaves a vote on 13 alternatives, all to be considered in parallel:

1. Accept Plan Harder
   - With Hole 7 left as it is
   - With Hole 13 left as it is
   - With both Holes 7 and 13 left as they are
   - As presented
2. Accept Plan Easier
   - With Hole 7 left as it is
   - With Hole 13 left as it is
   - With both Holes 7 and 13 left as they are
   - As presented
3. Change the front 9 to Plan Easier and the back 9 to Plan Harder
   - With Hole 7 left as it is
   - With Hole 13 left as it is
   - With both Holes 7 and 13 left as they are
   - As presented
4. Make no changes.

Fortunately, in this parallel universe, the club membership had previously reviewed its bylaws and, in so doing, had adopted majority judgment as the preferred voting rule. It can simultaneously consider these 13 different ideas, using the following language:

In the best interests of the Shady Grove Country Club, I believe that this plan is:
A = Excellent; B = Very Good; C = Good; D = Fair; E = Poor; and F = Unacceptable.

Under the club's new bylaws, the voters can also use "plus" or "minus" modifiers on the grades, just as in academic grades in high school and college. The members know from previous experience that they can use the same grade more than once and that they need not use all grades.

After discussion ends, the president calls a 10-minute recess and the secretary types up and prints ballots with the 13 options carefully specified, with the language of what each letter grade means precisely written on each ballot. The members return and are given the ballots, which they fill out in just a few minutes. The last ballot is handed to the secretary four minutes after voting begins. About two-thirds of the members use "plus" or "minus" modifiers on their grades, mostly to discriminate between the plans that they like but do not consider to be quite equal. The course revision committee chair tries to submit two ballots, but the deception is detected.

The president calls another recess. With the help of the treasurer, who reads each ballot, the secretary enters the grades into a previously prepared spreadsheet, taking approximately 12 minutes. The spreadsheet counts the number of specific grades assigned to each choice and automatically reports the median grades. With 47 voters, a typical grade summary might look like this for two of the choices:

	A⁺	A	A⁻	B⁺	B	B⁻	C⁺	C	C⁻	D⁺	D	D⁻	E⁺	E⁻	F	Total	Median
Choice 1	6	4	5	7	6	3	3	2	2	3	2	2	1	0	1	47	B
Choice 2	3	5	3	4	6	2	1	4	6	2	3	1	3	3	1	47	C⁺

(Eleven other candidates would follow in the actual vote tabulation.)

With 47 voters, a majority requires 24 votes, so the median just adds up from the left until the total reaches or exceeds 24. For

Choice 1, the total is 28 when B is included, making B the median grade. For Choice 2, the total doesn't reach 24 until you get to the C⁺ grade, which becomes that candidate's median grade. The same is done for each option.

The results are reported back to the president, who announces them. The secretary's report reads:

> The winning choice is "Plan Easier with Hole 13 left intact," with a median grade of B⁺. In second place is to accept "Plan Easier with Holes 7 and 13 left intact," with a median grade of B. "Make no changes" received a median grade of C. The minutes of the meeting will publish the entire summary and list of grades.

The members happily retire to the bar for drinks, with the exception of the course revision committee chair, who sulks off home and gets drunk alone. His favored "Plan Harder with Hole 7 left intact" received a median grade of D⁺.

The president makes a mental note to find a way to automate the voting for future events to avoid the second delay, a task that the club treasurer accomplishes in just a few hours with the assistance of her teenage daughter and son.

We do not have to rely on imaginary debates at imaginary organizations to observe the consequences of limited forms of debate. Consider "Brexit" as a real-world example. It all began with a simple yes/no vote and a debate around that concept.

In 2016, a public vote for the United Kingdom to leave the European Union unexpectedly won. The voter turnout for Brexit was relatively low, possibly because polls had shown the referendum would be defeated anyway. The underlying issues were far more complex but, in the first instance, appeared to center on control of immigration, which involved both immigrants' access to generous public benefits in the United Kingdom and effects on job security as immigrants entered the labor force. Another key issue was the net "contribution" of the United Kingdom to fund the EU—about €13 billion annually in gross payments, with €4 billion in EU spending returned to the United Kingdom, for a net of about €9 billion annually. Finally, the perception of British sovereignty came into play as the EU's authority slowly expanded over member nations.

With the "remain" or "leave" vote thrust upon the citizenry—and apparently with many not voting because they thought, wrongly, that their votes would not matter—the Brexit issue became paramount in British politics for years. The prime minister at the time, David Cameron, resigned. His successor, Theresa May, eventually resigned in 2019. Boris Johnson was elected as the new prime minister but could not readily gain an agreement in the British parliament over the terms of exit. The result was a political version of reality TV.

Two new elections were called. Scotland threatened to secede from the United Kingdom in order to remain in the EU. Northern Ireland and Ireland—physically on the same geographic island but with different national alliances—found themselves confronting separate rules for travel and immigration. They even considered uniting as a single nation to resolve the complexity of the situation. The issue exposed vast rifts between the nations that form the United Kingdom: England and Wales (by about 53 percent each) voted to leave the EU, whereas Scotland and Northern Ireland voted to remain (by 62 and 56 percent, respectively). Whether these results will call for disbanding the United Kingdom in the future remains unknown. When the COVID-19 pandemic was ensnarling the world in 2020, the EU and United Kingdom were trying to negotiate treaties for international interaction when the United Kingdom officially leaves the EU. The consequences could reverberate for decades.

Little to none of the parliamentary maneuvering focused on the core question: Should the United Kingdom leave the EU, and if so, under what terms? The parliamentary rules and ballot structure turned the entire debate into political weaponry rather than focusing on the real issues at hand. Voters were never given the opportunity to evaluate different options—just a simplistic "remain" or "leave."

Imagine how things might have gone differently if voters initially had the option to express the importance of these other issues with such choices as these (instead of just "remain" or "leave"):

1. Remain under all circumstances.
2. Remain only if the United Kingdom's net financial contribution can be reduced substantially.
3. Remain only if the United Kingdom can substantially increase its control over immigration.
4. Leave unless both (2) and (3) are met.
5. Leave under all circumstances.

Other "remain" conditions could also be expressed similarly. Voters could express their enthusiasm for each choice independently, thus signaling their true concerns to their elected representatives. With something like a majority-judgment vote or a range vote with a clearly identified vocabulary (making it the same as majority judgment but using numbers instead of letters), U.K. leaders would have had a much clearer pathway to decision instead of the four-year turmoil that followed the remain/leave vote.

Things could have been worse. To its credit, the EU had a process defining how "states" could leave the "union." In an earlier era, failure to define such a process led to the Civil War in the United States, so the Confederate States and the Union fought a war that killed, by modern estimates, 750,000 people, amounting to about one out of every 10 White males in the country at the time. In the Civil War, a key issue also involved migration—Could slaves escape to the North and be freed from their slavery? More U.S. residents died in that war than in all subsequent foreign wars involving the country, including World Wars I and II, the Korean War, the Vietnam War, and the Gulf War combined. At least the United Kingdom didn't have to go to war against the EU to "Brexit." In many ways, the United States still experiences the consequences of the Civil War.

In addition to these two issues considered, a third way to mitigate strategic voting is to be cognizant of framing effects. The way questions are posed often shapes the answer. In a thought-provoking experiment at University College in London, subjects were told to imagine that they'd be given some specific amount of money, say £50, and then a wheel of chance would be spun (with a visual indication of the probabilities of stopping on black or white). If the wheel of fortune stopped on white (with a 40 percent probability), they got to keep the entire £50. If it landed on black, the subjects got nothing. The expected value of the bet is of course £20. The participants were asked whether they wanted to participate in the wheel-of-fortune gamble or get some sure amount. The "rationality" test consisted of how often each subject gave the answer that a rational person would give. If asked to compare a gamble of £50 (with 40 percent probability) to a sure amount of £20, they should be more or less indifferent. But if the sure amount is £25, they should opt more for the sure amount than the gamble, and if the sure amount is £15 they should opt more for the gamble.

The key to the experiment was how the gamble was described. With the £50 starting point, the subjects were told that they could either go for the gamble (they'd get the entire £50 if it lands on white, nothing if it lands on black) or one of two other statements:

KEEP: You get to keep £20 of the £50 you got initially (with certainty).

LOSE: You will lose £30 of the £50 you got initially.

The amounts kept (e.g., £20 and £30) varied some during the experiment, sometimes giving the sure thing a higher value than the expected value and sometimes the reverse. A wholly rational person makes the same choice always, no matter which way the KEEP or LOSE framing is specified. If the gamble has higher expected value than the sure thing, they should go for the gamble, and vice versa.

But of course we don't always behave that way. These were the occasions when the subjects' choices conformed to the "frame." The opposite occurs when the choices don't conform to the frame. Every subject in the experiment responded to some extent to the framing—some more, some less. The framing effect says that the subjects will be more likely to accept the sure thing in the KEEP version and to prefer the gamble in the LOSE version.

This all hinges on the framing question of whether people "own" something or not. Their willingness to accept a risk requires a higher payoff when the risk involves losing something they possess. In the KEEP/LOSE experiment, the subjects "owned" the £50, so they would be more inclined to KEEP the "sure thing" rather than accept the gamble. So even a more-lucrative gamble (worth £25 expected value vs. a £20 sure thing) would be turned down more often in the

KEEP scenario. Behavior tilted in the other direction when given the LOSE framing.

Brain imaging showed that when people made intuitive judgments, they followed the predictions of the framing model, and when they made carefully calculated analyses, the "computing" parts of their brain got involved. In essence, people dislike risk when considering potential losses (like fire, theft, and health care costs) but prefer risks when thinking about things that can improve their well-being (like lotteries).

The fourth way to counter strategic voting is to think carefully about facilitators. We all know how strong personalities can drive group decisions. Much of the work published in the realm of decision analysis ignores the important role of the facilitator in affecting the decision. An early leader in work related to group decision-making was the RAND Corporation, established by the U.S. Air Force after World War II. RAND's key work for decades focused on improving decision-making in settings with imperfect information and different beliefs about the best way to proceed in environments with significant uncertainty (e.g., about how opponents in a military strife might react).

One of the key ideas from this early work at RAND focused on the power that influential individuals had over group decisions and ways to neutralize that power in order to reach unbiased decisions. RAND developed a technique called the Delphi, which has subsequently been modified in numerous ways. The technique was designed to elicit predictions of probabilities of various events and similar "hard" numbers. In the Delphi process, panels of experts are assembled, but never in the same room. Each expert is both anonymous and isolated from the others. All information is transmitted through a moderator who strips the information of identifying information. Thus, for example, a message saying "In my experience as a battlefield commander at Valley Forge in 1777–78, I know that the British soldiers only carried enough powder for 10 shots before they would change to bayonets." This would readily identify this "expert" as General George Washington, so the information about Valley Forge would be stripped from the message before others saw it.

As a hypothetical example, suppose a Delphi group were asked to estimate the number of cell-phones in Bangladesh. An initial round of estimates might ensue. Then somebody might ask, "What's the population of Bangladesh?" With an estimate of 150 to 175 million (the correct number is 165 million in 2020), somebody else might say, "I think there are about as many cellphones as there are humans in the world, so one cellphone per capita." (The correct number is a 96 percent connection rate.) Somebody else asks, "What about India?"

"I think India should be about at the world average, balancing income, landline system breadth, and how much technology they have."

"OK, if Bangladesh is like India, then there's about 150 million cellphones in Bangladesh." (The correct number is a bit over 158 million cellphones.)

However, if the question focused on Ethiopia, a very different reference would be needed. The connection rate in Ethiopia is only about 20 percent, one of the lowest in the world. Hong Kong has the highest rate in the world, with 2.57 cellphones per person. Cuba and North Korea are lowest at about 12 percent.

The point of this process is to isolate people from the effect of personality or identity of the source of information so that the content, rather than the source of expertise, becomes the focus of attention. However, even in this carefully controlled system, the role of the "moderator" looms large, since that person can alter or filter all comments going back to the group. But the emphasis of the Delphi process on anonymity is telling. The RAND researchers who developed this approach were very concerned about the overriding influence of strong personalities.

The point of this discussion is not simply to be aware of bias from facilitators. Sometimes facilitators are essential to good processes. The importance of powerful voices extends also to other people who can be put onto an agenda for strategic reasons. These include "keynote" speakers to kick off an event, summary speakers to end an event, and discussants in some formal settings. Every choice of speakers in key positions creates an opportunity to magnify that person's voice and hence point of view.

The four ways we explored are some approaches to mitigate the effects of strategic voting. We may never be able to eliminate strategic voting, but being aware of them is a start. We may learn to use them to our advantage or learn to detect when they are being used against us and our organizations. We should remember that the choice of voting methods that minimize the effects of strategic voting can also help. Voting methods that add up points (scoring, Borda count, approval voting, among others) are more prone to strategic voting than median-based methods (trimmed mean averages, majority-judgment grading). How important it is to deal with strategic voting will vary across settings and occasions.

## Chapter 6

Poll Vaulting

We now bring together some real examples from the world of social choice and systems analyses to illustrate their consequences. We use two examples from U.S. presidential elections. A third example focuses on a vote by 11 people that transformed the wine industry. The fourth example examines agenda manipulation, harking back to framing and sequencing of decisions.

Perhaps the most prominent failure of "majority rule" with three or more candidates was during the 2000 election for the U.S. president. Al Gore narrowly won the popular vote, but the United States uses a complicated system where each state's voters elect people who cast the Electoral College votes. Framers of the U.S. Constitution deliberately created this method to prevent large states from overwhelming small states. We focus not on the issue of Electoral College versus popular votes; rather, we discuss Florida's voting system, which focuses us on the appropriateness of majority rule when multiple candidates compete. The U.S. Constitution grants rights to the states to establish their own voting rules. Florida's rules give their entire 25 Electoral College votes to the plurality winner—the one

Making Better Choices. Charles E. Phelps and Guru Madhavan, Oxford University Press (2021). © Oxford University Press. DOI: 10.1093/oso/9780190871147.003.0006

with the greatest number of votes, even if not a majority. Here are the final Florida votes (5,963,110 total) from the 2000 election:

George W. Bush	2,912,790	(48.847 percent)	> 537-vote difference
Al Gore	2,912,253	(48.838 percent)	
Ralph Nader	97,488	(1.659 percent)	
All others	40,579	(0.35 percent each)	

In a runoff election, presumably (as guided by exit-poll verification) almost all of Nader's votes would have gone to Gore (had they voted in a runoff), and the remainder would have been split about evenly. If everybody voted in a runoff, Gore would have won by about 50.8 percent to 49.2 percent—a convincing victory. We will explore in this chapter how other voting systems would handle this same situation.

The same could have occurred in New Hampshire (four Electoral College votes). Bush won the Electoral College with 271 votes (one more than the needed majority) to 266 for Gore. (One elector did not cast a vote, so the total was only 537). Had New Hampshire's vote switched, Gore would have had the smallest possible winning majority of 270 votes to Bush's 267. In the final New Hampshire tally, we see 273,559 for Bush (48.1 percent), 266,348 for Gore (46.8 percent), 22,198 for Nader (3.7 percent), and other fringe candidates gathering about 1 percent. With Bush's edge over Gore at 7,211 votes, and the same presumption about Nader voters' choices in a runoff, Gore would have won a New Hampshire runoff as well. Either state's votes would have changed the Electoral College outcome and hence the choice of U.S. president.

We can extend this discussion to think about voting rules other than runoff elections. For simplicity, let's ignore the fringe candidates and focus on Bush, Nader, and Gore. The Florida voter profile is assumed to look like this (based on polls at the time of the election):

Bush > Gore > Nader	2,912,790 voters
Gore > Nader > Bush	2,912,253 voters
Nader > Gore > Bush	97,488 voters

Presuming that Nader's votes went to Gore almost exclusively, Gore would readily win a simple runoff election. But what about other voting rules? First, does a Condorcet winner exist? Given a presumed preference ranking of Nader > Gore > Bush for the 97,488 voters who listed Nader on their ballot, then a Condorcet vote would include Nader voters for Gore when comparing him versus Bush, and Gore would be the Condorcet winner (defeating both Bush and Nader).

How about the Borda-dependent methods, including straight Borda counts and elimination methods of Nanson and Baldwin? We have the following plausible Borda counts for Florida:

Bush	$2,912,790 \times 2 = 5,825,580$
Gore	$2,912,235 \times 2 + (2,912,790 + 97,488)*1 = 8,834,784$
Nader	$97,488*2 + 2,912,235*1 = 3,097,211$

Gore wins on a straight Borda count. Turning to the methods that eliminate candidates based on Borda counts, the average Borda count is 5,992,531, which eliminates both Bush and Nader, leaving Gore the winner. The Baldwin method drops Nader with the lowest Borda count, transferring his votes to Gore, who wins.

Similarly, in instant runoff voting, Nader would be dropped in the first round (fewest first-place votes), but votes for him would transfer to Gore, making him victorious. Extending this analysis, this presumptive Florida voting profile reveals an important difference between instant runoff and Coombs voting rules. Instant runoff drops the candidate with the fewest first-place votes (here, Nader). Coombs voting drops the candidate with the most last-place votes (here, Bush). Thus, if Florida had used a rank-order ballot and Coombs voting rules, Bush would have been eliminated first, then in the runoff using rank-order ballots, Gore would have defeated Nader. The outcome is the same—Gore winning with Coombs voting rules—but the process differs.

This highlights another issue—the incentives for sincere voting versus strategic voting. With instant runoff, people who truly favored Nader might well rank Gore ahead of Nader in first place on their ballot to ensure that Gore was not dropped in the first round. With Coombs voting, they could safely reveal their true preferences (Nader > Gore > Bush) without harming Gore's chances of election.

Gore would similarly win with Bucklin rule since Gore would rank first or second on a majority of the ballots. Thus we can see that in every voting rule considered until this point except plurality voting—the rule actually used in Florida and New Hampshire—Gore would have won Florida and New Hampshire, and hence the U.S. presidency.

Remember that we don't actually have rank-order lists of the voters' preferences in Florida or New Hampshire. It is entirely plausible that some of those who voted for Gore in the actual election truly preferred Nader but did not want to "waste" their vote.[1] Suppose that there are actually (say) 1 million of such people, and they could express that on a rank-order list showing Nader > Gore > Bush. That would not change the outcome for any of the rank-order methods discussed here. In fact, Nader would actually have to surpass Gore in #1 rankings to alter the election outcome using these rank-order–based systems. Had that happened, the Electoral College would not have been able to declare a winner and the U.S. House of Representatives would have selected the U.S. president.

A similar and more precarious vote occurred in 1992 when Bill Clinton won the U.S. presidency over George H. W. Bush and Ross Perot. Looking just at the popular vote, Clinton got 43 percent, Bush got 37.4 percent, and Perot got 18.9 percent. How would Perot's voters spread out with different voting rules? If Perot voters' second preferred choice split two to one in favor of Bush, the latter most likely would have won the election using most (if not all) of the election rules discussed earlier.

Exit polls at the time show that if Perot had not run, 38 percent of those who voted for him would have voted each for Bush and Clinton and 24 percent would simply not have voted. This result might not have altered the outcome. But this does not tell us how they would have marked rank-order ballots, or what their participation rates might have been if they had rank-order ballots. So, the

---

1. The reverse could also be true: Some may have voted for Nader even when their preferred choice was Gore. Why? Because U.S. election rules specify that once a political party receives 5 percent of the total vote, it becomes eligible for federal campaign funds in future years. So some voters may have switched to Nader from Gore to help boost the Green Party's total to 5 percent. This different incentive to voters further confuses our understanding of what the "vote for one" ballots might have become in a rank-order or runoff ballot system.

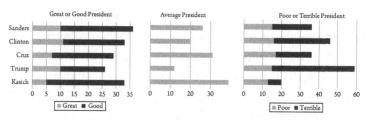

Figure 6.1. Grades on 2016 U.S. presidential candidates

actual outcome using different ballots and vote-counting rules is necessarily a mystery.

The point here is not to say that Nader in 2000 or Perot in 1992 should not have run, nor that their candidacies "gave" the election to Bush and Clinton in respective years. The key point is that different voting rules lead to different outcomes.

We have no way to tell what would have happened in Florida or New Hampshire had majority judgment been in use rather than plurality voting. However, there is a way to reconstruct what might have occurred in the U.S. presidential race in 2016 with different voting rules. Michel Balinski and Rida Laraki constructed a majority-judgment ballot from the 2016 polls led by the Pew Foundation.[2] This ballot, whose results are shown in Figure 6.1, showed "word grades" for leading presidential candidates of both parties: great, good, average, poor, terrible, and never heard of.

Counting "never heard of" as worse than "terrible," all of these candidates had a median grade of "average" except Trump, who received a median grade of "poor," and barely above "terrible." Using

---

2. This Pew poll was quite similar to the Q-score, a popularity rating used extensively in advertising and marketing, but with a different and perhaps less informative way of using the data than the majority-judgment process itself.

the majority-judgment rules, Kasich got a grade of "average plus" and all others except Trump got "average minus," and the overall candidate ranking would have been Kasich > Sanders > Cruz > Clinton > Trump. Removing "never heard of" from the list actually raises Kasich's standing compared with the others, but no candidate has either their majority grade or their modified majority grade altered by this removal.

In the actual election, we ended up choosing between the lowest-graded pair of the five final candidates (Trump and Clinton), which may explain some of the weak voter turnout issues that may well have determined the final Electoral College results. If the highly graded candidates from each party were put against each other in a direct comparison, the Republican candidate would have been Kasich and the Democratic candidate would have been Sanders. Based on these voter evaluations, Kasich would have beaten Sanders and would have been elected U.S. president. By comparison, in the actual U.S. primary election process, Kasich fell far behind Trump, Clinton surpassed Sanders, and Trump eventually defeated Clinton in the Electoral College, although both the Pew data using majority judgment and the actual popular vote count in 2016 clearly put Clinton ahead of Trump.

These data also illuminate the difference between instant runoff and Coombs voting. They both use rank-order lists and remove candidates with different rules. Instant runoff voting removes the candidate with the fewest first-place votes. In this table, that's Kasich, where only 5 percent ranked him as likely to be a "great" president, with Cruz eliminated second by that sort of rule.

In contrast, the Coombs rule drops candidates who have the most last-place rankings. From the table in Figure 6.1 (ignoring "never heard of"), Trump has by far the most last-place scores (44 percent listed him as likely to be a "terrible" president). Clinton is next worst by that standard. Both Trump and Clinton are highly polarizing candidates, at the top of the list in terms of people who think they would make a "great president" and also with the highest percentage of people who think they would make a "terrible president." So Trump and Clinton would fare better in instant runoff voting, whereas Kasich would do better with Coombs voting.

As with majority judgment, we cannot reconstruct a true "approval ballot" for Florida or New Hampshire in 2000, but we can approximate one for the 2016 election from the Pew survey data. Suppose that voters approve of all candidates rated "average" or above. This gives the resulting approval data:

	Approve	Not Approve
Kasich	72 percent	28 percent
Sanders	62 percent	38 percent
Cruz	59 percent	41 percent
Clinton	53 percent	47 percent
Trump	38 percent	62 percent

Based on these estimates, the final election (as with majority judgment) would have pitted Kasich against Sanders, with Kasich winning the general election. Instead, we had the option to choose between candidates with presumed approval ratings of 53 percent and 38 percent, and the lesser of those won the Electoral College vote. But this all depends on converting "average or better" grades to approval. It can change if "approve" has a different meaning. If the cutoff for approve was "good" instead of "average," the approval ratings would look like this:

	Approve
Sanders	36
Clinton	33
Kasich	33
Cruz	29
Trump	26

In other words, if approval is defined differently, Sanders (not Kasich) would win the election. This shows the volatility of approval voting when the meaning of the word is not defined carefully. And obviously, if different voters have different meanings for the word "approval," adding up their votes makes no sense. Still worse (and harder to understand the results) would be a case where some of the voters (in effect) used "average" and others used "good" as their cutoffs for defining "approve." Then we would be adding together numbers that had different meanings, just like adding together Fahrenheit and Celsius temperature readings.

Yet another real-life example of how voting methods crucially matter comes from the Judgment of Paris. This famous event was organized in 1976 by Steven Spurrier, an Englishman operating a high-end wine shop in Paris. The 2007 movie *Bottle Shock* loosely follows these events. The event pitted some newly emerging American wines (chosen by Spurrier) against top French wines, which at the time were deemed by the French as the best in the world. Eleven judges (nine French wine experts, one American wine expert, and Spurrier himself) tasted 10 wines: six American and four French for both the reds (Cabernet Sauvignon) and whites (Chardonnay). The tastings were blind, using numbered glasses. Judges were instructed to score 0 to 20 points to each.

But it was very important to the French that an American wine *not* win. Then the unimaginable happened: American wines won both the red wine and white wine competitions. The French newspapers completely ignored the results and then sneered at the process. Even Spurrier was shocked. One of the nine French judges tried to withdraw her ballot after hearing the results (she had ranked Stags' Leap, the winner, as #1 and otherwise low-ranking Mayacamas as #2). In the reds, Napa Valley's Stags' Leap just edged out Château Mouton Rothschild (one of the most famous of all French wine names) with average scores of 14.14 and 14.09. France took the next three slots, and the last four of the 10 reds in average score (see Table 6.1 for rankings) were all from California. Napa's Chateau Montelena took first place, and other Napa Valley Chardonnay wines garnered three of the first four slots in the final ranking of the whites.

The results transformed the global wine industry. They accelerated the commercial success not only of wines from Napa Valley but also those from Oregon, Washington, Australia, New Zealand, Spain, South Africa, Chile, and Argentina. All these regions are now successful producers of wines, some of them of very high quality. Famed wine critic Robert Parker said in 2001: "The Paris Tasting destroyed the myth of French supremacy and marked the democratization of the wine world. It was a watershed in the history of wine."

Clearly, this was a very important vote, even if taken only by 11 people. Did the method of scoring the wines matter?

We can begin by assessing these scores as they were intended— as scoring on a scale of 0 to 20. Let's first ask whether the judges were using the same scale—an essential question before taking averages or totals of their scores, and to make those scores meaningful.

Table 6.1 ACTUAL SCORES FOR PARIS WINE-TASTING EVENT IN 1976

Judge	Stags' Leap*	Mouton–Rothschild	Montrose	Haut Brion	Ridge*	Leoville	Heitz*	Clos du Val*	Mayacamas*	Freemark Abbey*
Judge										
Brejoux	14	16	12	17	13	10	12	14	7	5
de Villaine	15	14	16	15	9	10	7	5	12	7
Dovaz	10	15	11	12	12	10	11	11	8	15
Gallagher	14	15	14	12	16	14	17	13	9	15
Kahn	15	12	12	12	7	12	2	2	13	5
Dubois-Millot	16	16	17	13.5	7	11	8	9	9.5	9
Oliver	14	12	14	10	12	12	10	10	14	8
Spurrier	14	14	14	8	14	12	13	11	9	13
Tari	13	11	14	14	17	12	15	13	12	14
Vanneque	16.5	16	11	17	15.5	8	10	16.5	3	6
Vrinat	14	14	15	15	11	12	9	7	13	7
Average	14.14	14.09	13.64	13.23	12.14	11.18	10.36	10.14	9.95	9.45

* United States, Napa Valley.

Table 6.1 summarizes the average scores and Table 6.2 shows the average, minimum, maximum, and (from those values) the range employed by the judges for the 10 red wines.

It is hard to believe that they had the same meanings for the numbers, although all were widely acclaimed wine experts and the French wines they compared were all famous French chateaux. Three of the judges, Oliver, Tari, and Spurrier, had ranges as narrow as 6 points from top to bottom, while Vanneque had a 14-point

Table 6.2 JUDGES' SCORING SUMMARIES

Judge	Average	Maximum	Minimum	Spread
Brejoux	12.00	17	5	12
de Villaine	11.00	16	5	11
Dovaz	11.50	15	8	7
Gallagher*	13.90	17	9	8
Kahn	9.20	15	2	13
Dubois-Millot	11.60	17	7	10
Oliver	1.60	14	8	6
Spurrier	12.20	14	8	6
Tari	13.50	17	11	6
Vanneque	11.95	17	3	14
Vrinat	11.70	15	7	8
Overall average	11.83			

*Patricia Gallagher, an American, was involved in organizing the tasting.

range (from 3 to 17 points). The big differences came in their use of low scores. The range of high scores was fairly compressed (from 14 to 17 points), but the range of low scores was much wider (from 2 to 11 points). All of the very low scores were assigned to U.S. wines, which may well have stemmed from a desire to downgrade wines they perceived as American and where they correctly identified the country of origin. The lowest score assigned by a French judge to a French wine was an 8. Kahn aggressively downgraded what she thought were the American wines, but she also assigned her two highest scores to American entrants. (Kahn was the judge who tried to retrieve her ballot when the results were announced.)

The average scores by judge ranged from 9.2 to 13.9 on the 20-point scale (46 percent and 70 percent of the maximum). The larger average (13.9) is 1.5 times the lower average (9.2). To put this in a slightly different light, think about measuring distances in miles or kilometers. A mile is 1.62 kilometers. Would we ever compute averages of distances that were a mixture of miles and kilometers? Yet here, the judges varied in their average score by a factor of 1.5, about the same as the miles/kilometers factor. Based on the average scores they submitted, their "measuring sticks" differed from one another about as much as miles and kilometers differ. This provides a vivid example of the problem of trying to add cardinal scores in voting rules. If different voters have different meanings for the scale of 0 to 20, adding or averaging them has no meaning. Further, the spreads between the highest and lowest scores ranged from 6 to 14, showing either that these tasters had very different meanings for the numbers or some had more sensitive palates than others. Indeed, the Paris scores clearly demonstrate that at least the low range of scores had very different meanings to these judges.

With a little bit of modification, we can use the reported numeric scores to see how other voting rules would choose the winner.

Remember that *the* winner was central to the entire affair, so let's concentrate on what winners emerge from different voting systems rather than the entire set of rankings.

We can use the judges' scores to reproduce a majority-judgment vote, computing the middlemost value and using majority-judgment hierarchical rules as necessary. We can also calculate rank-order lists using the scores, but since judges often assigned the same numeric score to different wines, we have to deal with ties, a small but fixable complication using rank-order lists. We can also determine vote-for-one outcome by using each judge's highest-scored wine as their choice if they had only one vote through plurality and runoff. And we can assess approval voting by specifying a rule using the scores to decide what each would approve of or not. The rule we chose for this example says the judge would approve any wine with a score above the average of all grades assigned by each specific voter. Using the cumulative voting method, we assigned 100 points in proportion to the points actually given by each judge to each wine. How do the wines compare using these different processes? The results from the use of different voting methods were as follows:

Approval voting	Château Mouton Rothschild
Black method (Condorcet + Borda)	Tie between Stags' Leap and Château Montrose
Borda count	Tie between Stags' Leap and Château Montrose
Bucklin voting	Stags' Leap
Condorcet rule	No winner
Coombs voting	Château Montrose
Cumulative voting	Stags' Leap
Majority judgment	Château Mouton Rothschild

Nanson voting	Château Montrose
Plurality	Château Montrose
Range voting (what was actually used)	Stags' Leap
Runoff of top two	Château Montrose

This is a mixed bag of who gets to declare victory, but these results make it very clear that the choice of voting method determined the outcome in the 1976 Judgment of Paris. Of the 12 voting methods listed here, nine give an unambiguous winner (there is no Condorcet winner and there are two ties). Of those nine voting methods, a French wine wins six times (Mouton Rothschild twice, Château Montrose four times) and an American wine (Stags' Leap) thrice. The Borda count produced a tie for first place between Stags' Leap and Château Montrose. Since Black's method uses the Borda count to choose the winner if the Condorcet method doesn't produce a winner, it also gives a tie between Stags' Leap and Château Montrose.

This may be small consolation to French wine enthusiasts. Subsequent tastings of the same wines (in later years, as they aged more) all showed the American wines winning more often—mostly in comparing the reds, since the whites soon were past their expected lifetimes. The first follow-on tasting (two years later, in San Francisco in 1978) had the top three red wines from the United States. Two retests appeared in 1986, a decade later, one at the French Culinary Institute (U.S. wines had the top two) and the other at the New York–based *Wine Spectator* (U.S. wines took the top six slots, so all six of the U.S. red wines beat all four of the French reds). Spurrier created a 30th-anniversary event (a long period of aging even for top red wines) in parallel events in Napa and London that included a number of the original judges. In this high-profile event, U.S. wines took the top five slots among the reds. Thus, the long-term superiority of the U.S. wines in repeated subsequent blind tastings shows that the original results may not have been a fluke. If anything, the narrow margin of victory for the Stags' Leap red wine understated the superiority of the U.S. wines.

The white wine votes in Paris showed a stronger dominance of the U.S. over the French wines, in contrast to the reds, where Stags' Leap barely won. Most of the top-rated red wines were French and the bottom four were all from the United States. The top white (Napa's Chateau Montelena) outdistanced the closest French competitor by almost half a point per voter (in contrast to the tiny gap in the reds), with two of the top three and four of the top six whites coming from California.

While the actual votes on the whites are not available (they are lost to the ages), the average scores show a significant gap between the #1 Chateau Montelena (a U.S. wine despite the French word in the name), with a score of 14.67 points, and the #2 French wine, at 14.06. This probably means that Chateau Montelena would win with most if not all of the voting methods discussed. Further, because white wines do not age very long, no opportunity arose for another vote in later years. So, only the original 1976 results matter.

Particularly given the importance to the French that a French wine take first place, to repeat, voting methods matter. For wines they matter, and for tight elections they matter all the more. Declaring a U.S. wine as "the winner" in the Judgment of Paris revolutionized the world wine market.

Whenever there is a complex bargaining situation between parties, agenda manipulation is to be expected. This happens regularly in federal and state legislatures across the United States, but seldom do these strategic agenda-manipulation efforts see the light of day. One event in a private flying club has been carefully documented by its perpetrators, which gives us a good real-life example of how it works.

The event involved a private airplane club in Pasadena in southern California where Charles Plott, an economist, and Michael Levine, an attorney, were charged with establishing an agenda for the next club meeting. They were to decide what fleet of aircraft to recommend to the club's board of directors. The primary purpose of the club was to make private aircraft available for rental on the most desirable terms possible. Plott and Levine both wished to have at least one (preferably two) twin-engine, six-seater aircraft for their larger family travels. They also wished to keep the operating costs of these larger aircraft as low as possible. They devised an agenda with a sequence of votes that led to recommendations for a fleet that closely matched their preferences, even though they knew that a head-to-head comparison of that fleet with other alternatives would never have been chosen.

Their goal was clear: to devise a set of votes that narrowed down the alternatives so that when the final vote came between two choices, their preferred fleet would win the yes/no vote. In their own words: "Each item on the agenda was designed to eliminate by majority vote some set of alternatives from further consideration."

The committee chair for whom the agenda was devised apparently understood the consequences of their proposed agenda. On four occasions, the chair proposed motions that would have defeated Plott and Levine's attempts to guide the final vote toward their preferred goal. Crucially, at the beginning of the meeting, the group voted to adopt the agenda proposed by Plott and Levine. On all four occasions, the chair was ruled out of order (they were apparently following *Robert's Rules of Order* carefully). They eventually got a "mixed" fleet with seven aircraft (the other alternative was six) and with two (not merely one) six-seaters. The latter was made feasible by a sequence of votes that prevented a single but more expensive six-seater in a six-plane fleet.

A bit of parliamentary magic was necessary to get their most-preferred fleet. They had presumed (and later verified with post-election polling data from the members) that the most-preferred fleet would have consisted of six smaller (four-seat) planes with no six-seaters at all. Seeking to eliminate that choice, they proposed a specific alternative—(A) six planes or (B) seven planes—early in the voting process. They premised this on the notion that at least one six-seater would be more acceptable if the fleet contained seven aircraft (not six). It turned out that once they had chosen a seven-plane fleet, nobody had any interest in an unmixed fleet, so they moved directly to the secondary question: How many six-seaters, one or two?

Plott and Levine preferred two six-seaters for two reasons. First, it increased the chances of having one available when they wanted to rent it. Second, it would steer the board toward the less expensive of the two six-seat alternatives, which was also their preference. So they set the agenda first to choose between one or two six-seaters (two was the choice), and then the choice of the lower-cost types moved easily. They also strategically placed a cost-related question at the end of the agenda: How elaborate should the aircraft's avionic equipment be? In setting the agenda sequence, they had concluded that an earlier vote to equip the planes elaborately would scare members away from a seven-plane fleet, and hence increase the risk that no six-seaters would be included.

Plott and Levine had some understanding of the group preferences from previous meetings, but their knowledge was incomplete. This highlights one of the key issues about not only agenda-setting but also strategic voting in general—it's all for naught if you know nothing about other voters' preferences. If you are totally "blind" to how other voters feel, your best alternative is to vote sincerely.

In that sense, of course, public opinion polling prior to elections induces strategic voting. In Australia, where rank-order ballots and single-transferable-vote rules are used to elect members of their parliament, political parties regularly publish cards with recommended, rank-ordered voting strategies. This is to maximize their chances of obtaining majority control of the parliament and hence the ability to choose their prime minister without needing the compromise of coalition formation.

These real examples show how different voting methods to score various options matter. They matter even more when performance relies on multiple attributes that need to be carefully considered and blended. The first two examples showed this for voting for the U.S. presidency. The Paris wine example mostly highlighted how different voting systems work, but it also brings to mind the difference between having a single summary score (as in the 0-to-20 scale for each wine) versus multi-attribute systems analyses scores that would assess each component part of the wines independently, such as used in the Robert Parker, Australia, and UC Davis wine ratings.

The literature in social choice analysis and similarly the literature on multi-criteria decision analysis are unhelpful in choosing among the many approaches to voting (or systems analysis in general). The problem appears most starkly in social choice analysis, where the academic literature has evolved in a direction that almost guarantees no ideal and reliable voting model. That literature now has dozens of criteria that a good voting rule must meet. Proponents of every voting theory gleefully point out how competing models have failed this-or-that set of voting criteria—which, naturally, their favorite model passes. Thus, we have dueling criteria and also a morass of models.

In the next chapter, we extend our thinking on how people use the voting tools and not how these tools behave mathematically.

## Chapter 7

Human Factors of Democracy

Social choice analyses of various voting rules almost entirely ignore their human factors—a concept vital for engineering design. Accounting for human factors could help illuminate the possible effect of the tool on voter participation—the heart and soul of democracy. We focus on four relevant human factors: How expressive are the ballots? How easy is it to complete the ballot? How easy is it to understand the workings and results of the ballot? How enjoyable is the voting system to use?

With these guiding questions in mind, and leaning more heavily on the human factors than the mathematics of social choice criteria, we'll summarize our thoughts on the choice of voting systems for different environments (Appendix B provides a tabulation). The right choice may depend on the nature of an organization, the history of how votes have been conducted (familiarity may breed affection or contempt), and the importance of a rich vocabulary on the ballot.

After the 2016 presidential election, the *Washington Post* interviewed a number of non-voters to ask why they didn't vote. One

Making Better Choices. Charles E. Phelps and Guru Madhavan, Oxford University Press (2021). © Oxford University Press. DOI: 10.1093/oso/9780190871147.003.0007

respondent said: "Because I have no way to express my disgust and loathing for those in office or those who seek to replace them."

If we treat ballots as a way of conveying information from voters to the organization holding the election, having a larger vocabulary would certainly be useful. In this case, "more is better," but we do not know exactly how this works. The second issue—perhaps of greater importance—is the effect of the ability to "speak your mind" on voter participation—as exemplified by the respondent's frustration.

This part of the discussion focuses only on the human factors of the ballot, their vocabularies and expressivity, and for the following: (1) vote for one, (2) rank order, (3) approval voting, (4) scoring, (5) cumulative voting, and (6) letter grading. In some cases, a ballot type has only one algorithm supporting it. In other cases (notably rank-order lists), a large number of algorithms have been proposed to combine the ballots into a collective choice.

It's fairly easy to see that the six different types of ballots we have discussed provide very different degrees of ability for voters to actually express how they feel about the choices. Vote for one is the most limiting. Others offer greater expressiveness. Table 7.1 shows the formula for calculating the number of different expressions that voters have using these methods and then gives a specific example for five candidates ($k = 5$). The basic pattern holds for all choices except ranking and grading, where they reverse order at about 14 candidates (for six grades). However, by then, voters have so many ways of expressing themselves that it can't really matter.

The most common voting method in the United States and many other nations ("vote for one") is the weakest in its expressiveness. For k candidates, you have merely k different statements that

Table 7.1 VOTING METHODS AND POSSIBLE EXPRESSIONS

Inputs (voting method)	Calculating the number of expressions	Number of expressions with k = 5 candidates
Choose (select one candidate)	k	5
Approve (select all whom you approve)	2^k	32
Rank (1, 2, 3 ... k)	k!	120
Grade (A, B, C, D, E, F)	n^k for n grades (A, B, C, etc.)	7,776 with n = 6 grades (>1 million for m = 100)
Cumulative voting (spread 100 points)	m!/[k! * (m − k)!]	73.5 million for m = 100
Score (0 to 100)	m^k	10 billion if m = 99

you can make. For five candidates, you have five "words" you can say, and you are only permitted to say one thing on your ballot. Next weakest on the list (using the example of five candidates) is approval voting, with 32 possible expressions, followed by rank-order lists, resulting in 120 expressions. Except with one to three candidates, rank order is more expressive than approval voting but still very weak compared to others. Letter grading (with six majority judgment grades) gives almost 8,000 words for five candidates and over 1 million if you use "plus" or "minus" modifiers. Cumulative voting (dotmocracy) gives 73.5 million expressions, and scoring (with 100 points) yields 10 billion expressions.

In actual use, people almost entirely restrict themselves to scores at five-point intervals in scoring, so realistically there are merely only 21 scores leading to 4 million different statements for five candidates. If we assume the same about cumulative voting, the number of alternative statements for five candidates still exceeds 20,000.

To put these kinds of counts in perspective, a six-month-old infant has a vocabulary of about six words. A one-year-old reaches about 25 words. The typical 18-month-old child can say about 50 words but is beginning to understand two to three times that many. A smart dog understands (but cannot say) about 165 words. Border collies and poodles may understand up to 400 words. Thus, we can see that vote-for-one methods "dumb down" the ballot to the intelligence of a six-month-old infant. Both vote-for-one and approval voting methods provide ballot complexity within the capability of even a moderately intelligent dog. These ballots have little value as communication devices.

They are further limited by the rules of grammar that constrain how you can use the allowed words. Rank-order ballots only have k words and highly restricted "rules of grammar." Grading with six letters has a limited vocabulary but unrestricted use of the allowed words. Cumulative voting has a lot of words, but there are some restrictions on how you can use them (the total points have to sum to a specific total). Scoring allows a large vocabulary and has no restrictions on usage and hence gives the greatest number of possible expressions (see Table 7.1).

For any given number of candidates, rank-order ballots throw away or hide a lot of information that grading and numerical scoring allow. By definition, rank-order lists do not convey any information about the strength of preferences. They require that each rank be used and that each rank be used only once. This is a very restrictive notion.

Consider a voter list of candidates and their qualifications described in words:

Candidate Adrian:	Excellent to very good qualifications
Candidate Dale:	Very good to good qualifications
Candidate Casey:	Good, possibly very good qualifications
Candidate Kyle:	Poor
Candidate Rory:	Almost unacceptable
Candidate Morgan:	Unacceptable
Candidate Tracy:	Excellent

What might a voter "say" with various ballots?

Vote for one:	Tracy
Approval voting:	Approve Tracy, Adrian, Dale, and Casey ("good" and better) or Approve Tracy, Adrian ("very good" and better)
Rank order:	Tracy > Adrian > Dale > Casey > Kyle > Rory > Morgan
Majority judgment:	Tracy = A; Adrian = A-; Dale = B-; Casey = C+; Kyle = E; Rory = F+; Morgan = F
Scoring:	Tracy 98; Adrian 90; Dale 80; Casey 75; Kyle 35; Rory 15; Morgan 0
Cumulative voting:	Tracy 30; Adrian 26; Dale 20; Casey 14; Kyle 8; Rory 2; Morgan 0

One can see how weak the vote-for-one and approval ballots are. They provide very little way for voters to express themselves. Rank-order ballots are slightly better, but one must use every rank, exactly once. Further, ranks do not distinguish degrees of difference at all. If the intent is to provide information, rank orders are also a bit constrained by their structure.

Ballots can have multiple purposes. If the only purpose is to choose a winner, narrowly expressive ballots can be used, despite their defects. The more eloquent ballots provide a much richer ability to distinguish nuances of evaluation and preference. Unfortunately, ballots with richer vocabularies may also open the door to more strategic manipulation. As recognized earlier, methods that use medians rather than means or totals are least susceptible to strategic voting, and majority judgment has the least risk.

Grading, dotmocracy, and scoring offer voters more expressivity than vote for one, approval voting, and ranking. Thus, for organizations seeking to learn the thoughts and sentiments of their voting constituencies, more expressive ballots might be an option to consider. By far the worst in expressivity is the vote-for-one approach—the most commonly used ballot in the world for political decisions.

Let's now consider the ease of use. Some ballots let the voter make one-by-one assessments of candidates. This may provide convenience to voters. In this regard, vote for one may seem the simplest, but the method requires a comparison of all candidates to pick "the best." In that sense, it's akin to ranking, only easier. Once the winner is chosen, the job is complete.

Approval voting and scoring seem relatively simple to use: One involves checking each candidate being approved, and the other involves assigning a number for each candidate, both one at a time. Approval voting requires deciding in advance what "approve" means, and scoring requires deciding what "0" and "100" mean. For example, does "100" mean the best of the current set of choices, the best the voter has ever seen, or the best imaginable?

Majority judgment is similar to scoring but involves the use of letters corresponding to the perceptions within the context. Each candidate is letter graded independently—a "gestalt" evaluation. A mock ballot at a French presidential election was carried out at actual voting places, and people completed their ballots quickly and easily. The average time people took was about one minute per ballot, or about five seconds per candidate, since the ballot had 12 candidates.

Rank ordering may pose some difficulties with breaking ties or not being able to exclude candidates from a rank in a list. For example, one might want to rank 1, 2, 5, 5, 6, 6, 7, 8 for eight candidates to show a big gap between the first two and the rest and also to show some ties. Rank ordering typically does not permit this. Another issue is that the complexity of rank-order ballots increases exponentially with the size of the candidate list. In the cases of approval voting, scoring, and majority judgment, complexity increases only linearly with more candidates—one decision per candidate.

If we go back to the origins of the Condorcet idea (and any rank ordering), it basically involves a pair-by-pair comparison. Suppose that there are five candidates, A, B, C, D, and E. First you compare A versus the other four candidates, then B with the remaining three, then C with the remaining two, then finally D and E. This results in $4 + 3 + 2 + 1 = 10$ comparisons. With six candidates, the total goes up to 15. With eight choices, the total is 28. In the French mock election, it was observed that almost 60 percent of the participants did not rank all 12 candidates; over half of them ranked six or fewer of the choices. In Australia, where rank-order ballots are regularly used, 95 percent of voters use preprinted ranking lists provided by their favorite political party. Ranking is hard to do with many choices.

Next on our human factors list, do people actually understand the voting algorithms and their effects? The academic field of social choice analysis has not been helpful in that regard mainly because it has become increasingly mathematical. It should come as no surprise that attempts to find "better" voting methods are also becoming more complex mathematically. This has arrived to the point of incomprehensibility to typical voters and even technically trained professionals. Here, we review a few of the major voting rules for the ease of understanding of both their working mechanism and their effects.

In plurality voting, the candidate with the most votes wins. This plurality method was notable in the reality TV show *Survivor*, where every third day, one member of each tribe was voted off the island. Had the producers of *Survivor* required pure majority votes or Condorcet winners, they would not necessarily have removed anybody every third day, as the series required. Because of the similarity with the usual majority voting (when only two candidates are on the ballot), it is sometimes difficult to tell it apart from plurality voting. But plurality voting can produce irrational results (as we saw in the 2000 U.S. presidential election) where rank reversal can change the outcome depending on third-party candidates. This may not be obvious to the average voter.

Runoff voting, although a bit less common than plurality voting, has the same benefits as in ease of understanding and the same pitfalls primarily due to rank reversal, *Survivor* used runoff voting in the few cases where more than one person received the same plurality vote.

Condorcet voting is easy to understand when described as the candidate who beats all others head to head wins. What's not well understood is that the method can fail to produce a winner, as seen when Condorcet cycles appear in games such as *Rock, Paper, Scissors*.

Borda count is used to rank collegiate sports teams on a weekly basis during the relevant sports season, as well as choosing most valuable players in many sporting settings. The problems of rank reversal and ease of manipulation and the weak ability for people to express their true feelings using rank-order lists are the main limitations. Additionally, Borda-related rules such as Nanson and Baldwin are also difficult to explain because they require reranking and rescoring once candidates are removed.

Instant runoff and Coombs are modestly difficult with their rules since (as with Nanson and Baldwin) they too require reranking of candidates, but without a Borda count. When a single candidate is removed, those below that candidate move up one position. That eventually brings up some votes "from below" that lead to a majority vote for one of the remaining candidates, but it's not necessarily the same as the person who had the lead in the first round (which, we suspect, will be what most people expect to happen). As seen in the 2000 Florida presidential election, Bush had the greatest number of votes (as plurality winner) before anybody was dropped but loses when Gore receives votes from those who had Nader listed first on their ballot. Since people are so accustomed to having a plurality vote win the election, this might be source of confusion. And as with Nanson and Baldwin voting, the real complexity arises when trying to understand the effects of rank reversal.

Bucklin voting should be easy to understand and is closely linked to people's innate sense of what majority rule means. Is there a majority winner? If not, does one exist for people's first *or* second choices? It may be easier to understand why Gore wins in Florida in

2000 using Bucklin rules than explaining why he wins using instant runoff or Coombs rules. Gore was the first or second choice of a clear majority.

With majority judgment, the middlemost concept is easy to understand and it evokes a majority ballot vote. Obviously more difficult are the rules when two or more candidates have the same median grade. This probably stands as the biggest obstacle to better understanding and expanding the use of majority judgment.

Cumulative voting is used more often than people realize, since it is a standard format for electing members for corporate boards. In California and the state of Washington, for corporations that are not publicly traded, cumulative voting is available by default to shareholders, even if not specified in the corporate bylaws. A difficulty often experienced is the need for cumulative voting to add up the votes to the proper number, an issue that vanishes with any electronic voting system. The less understood feature of cumulative voting is the potential for rank reversal, and the effect that as the number of votes allowed per ballot rises, so does the potential for strategic voting.

Scoring is usually simple to describe and is becoming increasingly common in use through the widely adopted "star" voting systems seeking customer feedback. These "star" systems are scoring methods with ranges typically of zero to four or zero to five stars. Particularly as the permitted scoring range increases, the opportunity for strategic voting expands.

The biggest issue about scoring is that adding up the votes makes no sense without a common language, an issue unlikely to be understood by voters, thereby putting some at a disadvantage in expressing their opinions. As noted earlier, voters might interpret 0s and 100s differently: Do they mean the worst and best in this slate of candidates or the worst and best candidates I can imagine? This will probably not be apparent to most voters, who will know what meaning they assign to the numbers and will probably assume that all other voters have the same meaning.

Approval voting might seem easy to describe, but in fact it is hard to comprehend in terms of "what happens" when one places an X next to each candidate approved. Indeed, voting theorists have trouble describing it in comparison to other voting methods. Is it scoring with a very narrow $(0, 1)$ range? (Yes.) Is it majority judgment with a limited number of grades? (Yes, only two grades, except that it adds scores rather than finding medians.) Is it like ranking? (Yes, sort of. It gives only two classes of ranks, the first and the last, and lumps candidates into one of those two boxes.)

Next, the phrase "enjoyment of use" in some sense combines ability to express oneself, ease of use, and ease of understanding how the voting tools works. The first two relate to the ballot itself, and the third relates to the algorithm used to create a group choice from individual votes. How much each of these issues matters and hence the consequent enjoyment may differ not only from person to person but also from occasion to occasion. In some settings, simply picking a winner is most important, so expressivity may have a secondary impact on enjoyment. But if the voter wishes to send a signal to those holding the election, then expressivity may loom large. Think of the non-voter quoted at the beginning of this chapter, who chose not to vote because the ballot gave no way to express their disdain for the proffered choices.

Having explored some human factors considerations, let's turn briefly to how scholars of social choice theory evaluate voting methods. Social choice theory focuses on formal criteria to evaluate different voting rules, often with complex mathematical arguments underlying the analysis. Literally dozens of such criteria exist.

Two problems emerge here that ultimately make the formal social choice literature relatively useless in the actual task of choosing a voting method. First, some of these criteria conflict with one another, and passing some criteria guarantees failing on others.

The second is a logical conundrum arising from the first: if a voting system can't pass all known criteria (and none does), then choosing a voting system becomes a matter of making trade-offs. Which is the lesser of various evils regarding a voting system? But to do that formally, one needs a voting system, say, to rank the importance or "goodness" of each of the criteria, which requires a vote. But that requires choosing a voting system. The process of trying to formally choose a voting rule itself becomes an endless loop. So the very logic of traditional social choice theory stands on a perilous cliff edge!

In practice, our reading of the literature on social choice theory is that proponents of one particular voting model seem to glorify the criteria that their own method passes and disparage those criteria that perceived competing systems fail. To make this a bit more concrete, let's look at two voting rules that both make sense on their own. One is called "join consistency," which was apparent in two recent U.S. presidential elections. To be join consistent, the outcome should not change if you combine two otherwise distinct voting districts. The other is the "Condorcet rule"—any candidate who beats all others head to head should win the election using rank-order ballots as the basis. It turns out that these rules are mutually inconsistent: No Condorcet voting system is join consistent, and no join-consistent model is Condorcet compliant.

The U.S. Electoral College is not join consistent. California casts 55 Electoral College votes for the state's winning candidate no matter whether that victory was by one or 1,000 or 1 million or 10 million votes. Adding up the actual vote counts—the "popular vote"—*is* join consistent. The inconsistency becomes more apparent when the Electoral College winner (Trump in 2016, Bush in 2000) did not receive a majority (or even a plurality) of the popular vote. It turns out that the only voting systems that are join consistent are those that add points (majority voting, Borda count, approval voting, and scoring).[1] The framers of the U.S. Constitution deliberately made the Electoral College not join consistent, fearing that the larger states could overrun the smaller states in a national majority-rules vote. At the extreme, it's possible to win the Electoral College with barely over one-quarter of the popular vote.[2]

Another of these arcane criteria is the "later no harm" rule, which says that in any election, a voter giving an additional ranking or positive rating to a less-preferred candidate cannot cause a more-preferred candidate to lose. That is to say, if you most-prefer candidate B, you can't harm B's chances of winning by up-ranking less-preferred choices A or C. But in the weird world of social choice theory, only two known voting method pass this test—runoff elections and instant runoff elections. But these voting methods have a highly dreaded outcome: They are not necessarily

---

1. Majority judgment is "grade consistent" in the sense that if two divisions (say, states) give the same grades to a candidate, then when joined together, they will also yield the same grade. But if, for example, one state has a median grade of B and another C, then combining them can alter the final outcome because one state's grade will be switched, either from B to C or from C to B.

2. To roughly describe the proof, suppose that Electoral College votes are approximately proportional to states' populations. Now suppose that in some states representing about half of the U.S. population, candidate A wins by the smallest possible majority. Now suppose that in the other half of the states, candidate B wins almost every vote. Then the winner of the Electoral College will have approximately one-quarter of the popular vote. This is an extreme example, of course, but it demonstrates how mismatches can occur between Electoral College and popular vote outcomes.

monotonic. "Monotonicity" simply means that in single-winner elections, no winner is harmed by being given a higher score or ranking, and no loser is helped by being given a lower score or a lower ranking. Monotonicity is an important rule on simple human factor grounds: It shouldn't be possible to harm a favored candidate by assigning a higher score—that badly violates the ease of understanding.

As discussed earlier, Arrow's impossibility theorem set the stage for this problem. Several modern voting methods have stepped aside from this paradox by allowing different types of ballots, such as approval voting, scoring, and grading. But these approaches introduce their own form of irrationality if voters possess different meanings for the scores or grades.

Readers should also be alert to the fact that score-based systems in general are not majoritarian, in the sense that they require majority approval of the electorate to win. "Majority wins" voting, by definition, does this, but at the cost of sometimes not producing a winner, and the same is true of all Condorcet-compliant voting methods. Other score-based systems can readily elect somebody who is not favored by a majority, just like plurality voting does with a vote-for-one ballot. The only voting rules we know of that require majority approval to win are "majority voting," majority judgment (where the majority agrees on the proper grade for each choice), and Bucklin voting (which fudges the definition of majority until one is reached). So if majority rules is an important concept in your organization, you have a limited set of choices—these three rules plus runoff and instant runoff voting, which present their own major problems. So now we are left with a puzzle. If no voting method can pass all of the known criteria, and indeed, if passing some of them guarantees failure on others, what possible good can these criteria bring?

As we have emphasized throughout this book, complex problems have multiple attributes, and each attribute can be perceived and valued differently. It is unrealistic to gain ground on every dimension of value with one candidate—such perfection may indeed be impossible. The same must remain true about choosing voting systems and rules for debate that accompany them. No voting system allows you to have it all in different measures of merit. Thus, there is no single voting system that is the best choice for every situation.

Epilogue

Vox Populi

We live in a world with the consequences of pretending that complex systems have binary answers. Having open discussions on these complexities is sometimes difficult, perhaps even painful. But awkwardness and pain on such issues merely represent hidden problems that inhibit our collective progress. Ignoring them or stifling discussion about them will not make them go away.

In *Making Better Choices* we have seen how the ways we choose affect how organizations make decisions. Two major branches of human thought have approached this issue, but from wholly distinct directions. Systems engineering has created a repertoire of ways for product and service design with performance improvement and new insights as defining features. However, systems engineering seldom deals with the question of how to do this in the context of a group of individuals acting as the decision-maker. A traditional systems engineering approach develops requirements of a desired, even acceptable, solution to the problem but doesn't necessarily consider where those come from. This is the realm of social choice.

Social choice analysis, however, takes the set of objects under consideration as given. It asks how to choose among a finite (usually

Making Better Choices. Charles E. Phelps and Guru Madhavan, Oxford University Press (2021). © Oxford University Press. DOI: 10.1093/oso/9780190871147.003.0008

small) set of alternatives. However, it does not create ways to improve upon the choices. The "candidates" emerge. As can be seen in recent U.S. presidential election seasons, early decisions on candidates are not that straightforward. The candidates' popularity sways more votes than does substance. Momentum dominates the portrait of success. In 2019, the Democratic Party had 20 declared candidates participating in early debates. In the 2016 election, the Republican Party had over a dozen on stage together in debates. Usual voting rules in U.S. elections are poorly equipped to deal with candidate slates of this size. It is very easy for somebody to emerge as the winner who is actually the favorite of only a small minority, particularly in "winner take all" votes.

Thus, social choice theory and systems engineering have operated in different spheres of enabling decisions. Our task in this book has been to marry them. We have also argued for voting systems that have a rich vocabulary, that are easy to use, and that are relatively immune to strategic voting if that is an issue. Range voting, majority judgment, and cumulative voting offer rich vocabularies. Vote-for-one ballots, approval ballots, and in most cases rank-order ballots have vocabularies that are significantly limited. Majority judgment also provides a common language for issues requiring a vote. On the downside of this, few people have any meaningful experience with majority judgment ballots. They are not difficult to use, as we have argued, but they are unfamiliar, both in use and in understanding how choices are made in the majority judgment system. Most people have high (and increasing) familiarity with range voting through the use of "star" voting, ubiquitous in modern societies. We see these in ratings of drivers (and passengers!) of shared-ride services, wine ratings, scores for products and sellers of services, and even ratings of the cleanliness of restrooms in airports.

To make expressive voting systems meaningful to our groups and ultimately our society, we also need to find ways to phase out rigid parliamentary procedures that control voting. These "rules of order" may be necessary when all complex problems are artificially broken into sequential binary votes, but once you have voting systems that allow evaluation of multiple options at the same time, the logjam can be broken. So we need to change both the way we vote and the way we think about, discuss, and debate options.

To do this, we need to adopt coherent, expressive voting systems that evaluate multiple options together and enable discussions that allow complex decisions to be discussed in their totality, not piece by piece, sequentially. This systems mindset will be essential to better design our democracies and evolve with them.

We cannot leave our discussion of democracy without considering the importance of diversity and inclusion. Diversity has more dimensions than membership. Many business organizations sell products to the general public. Understanding the views of that public about their product can be essential to survival. Having methods that allow people to express themselves clearly and eloquently can have great business and social value. Our earlier discussion about voting methods highlights how different various voting methods are in allowing communication about choices. Polling of constituents is the same as voting, except that no formal choice follows a polling exercise. It is advisory voting. And unless the poll asks all relevant stakeholders beyond a "representative sample," the results can be simple and misleading.

Lack of diversity has hindered many areas of scientific progress. Until very recently, scientific and engineering fields were almost exclusively the domains of White males. Just look at the lists of recipients of Nobel Prizes in physics, chemistry, medicine or physiology, and economics to see this. Only three women have ever won a Nobel Prize in physics (Marie Curie won both physics and chemistry prizes, and died of aplastic anemia, an autoimmune disease caused by radiation exposure from her own research). Five women have won prizes in chemistry and 12 in medicine or physiology. Only two women have been awarded the Nobel Prize in economics. Only one woman (from Iran) has ever won the comparable mathematics prize—the Fields Medal, which has been awarded 52 times.

These results primarily reflect the state of the world available to the prize committees. In 1900, about three-eighths of all college students were women. In the 1960s, medical schools overtly discriminated against women (and less overtly on the basis of race). In 1970, under 10 percent of medical students were women. In 2019, women held a slim majority of those slots for the very first time. In 2017, men still dominated science and engineering graduate programs (77 percent in engineering, 75 percent in mathematics and computer sciences, 66 percent in physical and earth sciences). Women outnumbered men in other fields such as public administration; health professions, including nursing and therapy specialties; education; and social and behavioral sciences.

In medical research, the bias against woman and underrepresented minorities appears in two ways. The awarding of research grants has long favored White males over others, in major part (most likely) due to the pool of applicants found in the nation's leading research universities. But the research regularly discriminated against women through enrollment as voluntary subjects, often focusing entirely on males. Not until 1987 did the National Institutes of Health establish guidelines requiring recruitment of women into clinical trials, and only in 1993 did Congress make inclusion of women and minority subjects an issue of law. One obvious effect of this earlier exclusionary practice was that fundamental biological differences between women and men (and how they react to different treatments) were ignored, often to the peril of women.

The problem starts not with colleges and universities but rather far earlier in the educational process. In Advanced Placement tests over recent years, *U.S. News and World Report* has noted, boys outnumber girls by more than four to one among computer science test-takers but by more than 2.5 to one on physics C tests, by nearly two to one in the more general physics B exam, and by nearly 1.5 to one on the calculus BC exam. Similar data show lower participation among African American and Latinx students. The "sorting out" of female and non-White students extends back into K-12 education, the specific point of origin remaining uncertain.

Fixing this issue will require a major overhaul of our K-12 educational systems, and perhaps even the scientific enterprise, throughout the United States. This may be not so much a matter of resources as one of attitude. Somehow, along the way from kindergarten through high school, and even during studies for advanced degrees, women and underrepresented minority students feel unwelcome in science and engineering-related fields. We do not pretend to know how to improve this situation, let alone identify mechanisms for racial reconciliation, but we believe it's important to identify these issues that affect democracies.

Unwarranted exclusionary practices cripple organizations of all types, and ultimately societies. Here we say "unwarranted" to acknowledge that some groups, such as religious bodies, are by nature exclusionary, admitting only those who profess agreement with the religious beliefs of the organization. The law of large numbers tells us this is true. The "best" of any skill increases with the size of the population sampled. Artificially shrinking a given population limits its skillset. This is true whether talking about students entering college or graduate programs, members in private clubs or policy committees or scientific academies or corporate boards, or the broader issue of polling some members of the general public about their views on any specific issue. Organizations that want to be the best that they can be cast wider nets. Going back to one of the chief tenets of systems engineering: We can't make everything better all at once, and we need to confront trade-offs. But good leaders and engaged citizens will help their organizations make better sense of these problems. We hope *Making Better Choices* makes a small contribution to this conversation and toward a much-needed change in choosing how to choose.

ACKNOWLEDGMENTS

Our first thanks go to the late Kenneth Arrow and Michel Balinski, both brilliant in their analytical power and concern for humanity. They offered us essential guidance, in person and through their writings, and have instilled in us ways of thinking that will endure.

We appreciate the insights and support of Rino Rappuoli, Alvin Roth, Harvey Fineberg, Rita Colwell, Victor Dzau, Bill Rouse, and Eswaran Subrahmanian, and the anonymous reviewers of the original concept as well as our draft manuscript.

We are grateful to the Department of Health and Human Services for funding our earlier work at the National Academies of Sciences, Engineering, and Medicine that eventually led to this book.

We salute our gem editor Abby Gross for her essential and enthusiastic guidance, and recognize the vital contributions of Katharine Pratt and colleagues at Oxford University Press in making this book into a reality.

Our final, and most important, thanks go to our spouses. Dale and Ramya inspire us every day in every way. As physicians, they care for people, alleviate their suffering, and improve their lives. We hope to produce a fraction of their impacts through our ideas.

APPENDIX A

Tie-Breaking in Majority Judgment

Majority judgment uses a set of hierarchical rules to determine winners with the same majority grade (that is, say, two or more have a grade of C). For any two candidates x and y with the same majority grade, define p_x and p_y respectively as the number of votes above the median. The winner is whoever has the larger p value. If they have the same p value, the number of grades below the median grade, called q_x and q_y is calculated. Whoever has the smaller q value wins. By definition, if their p values are the same and one has fewer q values, then that candidate has more grades at the median, since then m is the number of median grades and n is the total number of voters, $p + q + m = n$.

In rare situations, these rules fail to determine a winner, and one must use the basic process of eliminating votes one by one until one of the scales tips.

Here's a simple example:

Candidate x	ABCDE
Candidate y	AACDE

Here, both have the same median grade (C), the same p value (two above the median grade of C), and the same q value (two grades below the median grade of C). Here, the pure majority-judgment process first removes the median grade of C, leaving

Candidate x	ABDE
Candidate y	AADE

With an even number, the lower of the two middle grades is the median, which removes the D for both, leaving

Candidate x	ABE	(Median grade B)
Candidate y	AAE	(Median grade A)

Here, candidate y wins with a majority grade of A.

Now, let's consider the methods to get a full rank-order list when some of the candidates have the same median grade. This example uses the same going-to-dinner vote used in Chapter 4. In that example, Brazilian has a median grade of B and thus is the winner. The remaining three choices all have a median grade of C. The pure majority-judgment rule says to take away from each candidate, one at a time, a single median grade (median grades at this point are the least informative) and to proceed until the median grade of some candidate changes. Remember that when the number of grades is even, you pick the lower of the pair at the middle. Eventually, as the median grades change, you can tell which candidate is ahead. An example of this process is shown in this sequential removal of grades for the Delhi diner, which comes in second in this process.

First removal	A A A B B B B B **CC** C C C C E E E F	(18 grades)
Second removal	A A A B B B B B **C** C C C C E E E F	(17 grades)
Third removal	A A A B B B B **BC** C C C E E E F	(16 grades)
Fourth removal	A A A B B B B **B** C C C E E E F	(15 grades)

At the third removal, the median grade is the middle pair (BC) and hence remains as C, but at the fourth removal, the median grade becomes B. If you do the same thing for American and Chinese, you'll see that they both still have median grades of C after four removals, so Delhi moves ahead of them on the list. Continuing to remove grades would show that American would come in third place, and Chinese would be in last place.

This pure rule is tedious to do with a large number of voters, so this brings into play a useful shortcut called the "majority gauge" that produces the same winner as the majority-ranking process produces. Further, the only way the majority-judgment system creates a true tie is when the distribution of grades is identical for both candidates, a rare scenario with larger electorates.

This rule counts the number of grades above the median (p) and below it (q). You can read this directly from the table summarizing the grades. This creates a modified "median grade." If $p > q$, the modifier is plus, and if $p < q$, the modifier is minus. For our original table of grades for the four restaurants, we have

	Median Grade	Median	p	q	Modified Median
American	C	7	7	5	C⁺
Brazilian	B	4	7	8	B⁻
Chinese	C⁻	8	5	6	C⁻
Delhi	C	7	8	4	C⁺

Brazilian wins, of course, with the modified median B⁻ grade. This brings into play the first hierarchical way of resolving ties in median grades in the shortcut system. The modified grade of C⁻ for Chinese puts it in last place—more grades below the median than above. The comparable voting question here asks for each candidate: "If the grade were to be C⁺ or C⁻, which would you prefer?" The $p - q$ difference tells you which way such a vote would go.

How do we choose between American and Delhi? Since both have a C⁺ median grade, the shortcut method (majority gauge) next compares their p scores. Delhi get eight and American gets seven, so Delhi comes in second place, and we have a final ranking of Brazilian > Delhi > American > Chinese (or, in our previous shorthand, B > D > A > C). If the two tied candidates had minus modifiers (e.g., C⁻ instead of C⁺), we would have compared their q scores, with the smaller q score having the higher ranking.

A final comparison can come into play in the highly unlikely situation (with a large number of voters) that two tied candidates have the same median grade, the same modifier (say, plus for the moment), and the same total count above the median (p values). The next step would then ask how many q votes each candidate has (those below the median). With a plus modifier and a tie in p votes, the one with fewer q votes wins, since that candidate has more grades at the median (C in our example). A completely symmetric logic occurs if the modifier is minus instead of plus.

If two candidates have the same median grade, the same p values, and the same q values, then it must be true that they also have the same number of median grades. This means that the two are tied using the majority gauge. In a traditional majority-rule election, this can happen with an even number of voters who are exactly split in their preferences. In majority judgment it happens less often, because the mathematical requirements about dividing up the electorate are more stringent. In this case, one can turn to the pure majority-grade strategy to determine the winner.

How might the majority gauge fail to determine a winner when the full majority-judgment process would? The following example shows how this might happen (understanding that it will occur only very rarely with larger electorates). Suppose you have two candidates with the following grade distributions (with seven voters):

Candidate x	AABCCDE
Candidate y	ABBCCDE

Both have a median grade of C. Both have a modified median grade of C⁺, since for both, $p = 3$ and $q = 2$. Nothing in the simplified majority gauge system can distinguish one from the other, but candidate x clearly wins in the full majority-judgment process. This must happen since as you start removing median grades, eventually you get to the point where the median grade for candidate x is an A and for candidate y is a B.

This takes us back to the analogy of using a pair of scales of justice—one for each of two tied candidates with the same median grade. For each candidate, think of all of the median grades as stacked up on the pivot point. They have no effect on which way the scale tips (toward or away from the top grade). What determines that is the difference in count of votes on each side of the pivot point. That's what $p - q$ tells us. If $p > q$, then the scale tips toward the high grade, or away from it if $p < q$. That's the first step in the hierarchical rule—the plus or minus modifier.

Now suppose that both of the tied candidates have the same plus modifier. Next, we want to know which candidate's scale has tipped further toward the high grade. You do that by comparing which one of them has more votes above the median (compare their p values). The larger one wins. If they have the same p values, then you ask which has a smaller number of votes on the other side (their q values). The one with the smaller q tips more toward the high grade if they have the same p values. This happens, of course, because the winner has more votes at the median than the other candidate. (Remember that the vote total for all candidates is the same.) The process is entirely symmetric if both candidates started out with minus modifiers.

Note that in this vote-counting method, all votes above the median (or below it) count the same, no matter how far away they are from the median. If the median is C, then A and B count the same (above the median), and similarly D, E, and F (below the median) count the same. This is deliberate in majority judgment. They seek a grade that can get approval of the majority, but the intensity of preference doesn't matter in this type of vote (just as it doesn't in a simple majority-rule vote with two candidates). In that sense, it is not like a child's teeter-totter, where you have more leverage the further you are from the pivot point. It's more of a balance scale, where all of the votes on each tray count the same.

APPENDIX B

Comparison of Voting Tools

Voting Tool	Strengths	Weaknesses
Majority rule	– Works well with two choices	– Often fails if three or more choices
	– Easy to understand and conduct	– Weakest expressivity
Plurality	– Similar to majority rule	– Weakest expressivity
	– Easy to explain and conduct	– Risks rank reversal
		– Invites strategic voting
Runoff	– Similar to majority rule	– Major risk of rank reversal
	– Easy to explain	– Major risk of strategic voting
		– Not monotonic
Condorcet rule	– Analogous to majority rule	– Often fails to produce a winner
		– Risks rank reversal

APPENDIX B

Voting Tool	Strengths	Weaknesses
Bucklin voting	– Uses familiar rank-order list – As close to majority rule as one can get when using rank-order lists	– Risks rank reversal
Rank ordering* (*Borda count and methods that serially remove candidates, including Nanson, Coombs, Baldwin methods, and (the non-monotonic) instant runoff voting)	– Familiarity among voters	– Difficulty grows with the number of candidates – Risks rank reversal – Invites strategic voting – Each algorithm can produce a different winner
Approval voting	– Simple to explain and use	– Lack of common language complicates interpretation
Scoring	– Easy to use – Familiarity among voters – Allows full expression of relative values, allows ties and gaps	– High risk of strategic voting – Lack of common language complicates interpretation of scores
Cumulative voting	– Allows full expression of relative values, allows ties and gaps	– Some risk of rank reversal – Voting errors on ballots (if the totals don't sum to 100)
Majority judgment	– Allows full expression of relative values with ties and gaps – Less susceptible to strategic voting	– Voters have minimal experience – Tie-breaking methods may seem obscure

BIBLIOGRAPHY

Aldrich, John, André Blais, and Laura B. Stevenson. 2018. *Many Faces of Strategic Voting: Tactical Behavior in Electoral Systems Around the World*. Ann Arbor: University of Michigan Press.

Arrow, Kenneth J. 1950. "A Difficulty in the Concept of Social Welfare." *Journal of Political Economy* 58(4):328–46.

Baldwin, Joseph M. 1926. "The Technique of the Nanson Preferential Majority System of Election." *Proceedings of the Royal Society of Victoria* 39(1926):42–52.

Balinski, Michel, and Rida Laraki. 2007. "A Theory of Measuring, Electing, and Ranking." *Proceedings of the National Academy of Sciences* 104(21):8720–25.

Balinski, Michel, and Rida Laraki. 2011. *Majority Judgment: Measuring, Ranking, and Electing*. Cambridge, MA: MIT Press.

Balinski, Michel, and Rida Laraki. 2016. "Trump and Clinton Victorious: Proof That U.S. Voting System Doesn't Work." *The Conversation*, May 9.

Bertrand, Marianne, and Esther Duflo. 2016. *Field Experiments on Discrimination*. National Bureau of Economic Research Working Paper No. 22014.

Bertrand, Marianne, and Sendhil Mullainathan. 2004. "Are Emily and Greg More Employable Than Lakisha and Jamal? A Field Experiment on Labor Market Discrimination." *American Economic Review* 94(4):991–1013.

Copeland, A. H. 1951. *A Reasonable Social Welfare Function: Seminar on Applications of Mathematics to Social Sciences*. Ann Arbor: University of Michigan.

Dasgupta, Partha, and Eric Maskin. 2004. "The Fairest Vote of All." *Scientific American* 290(3):92–97.

DiStefano, Michael J., and Jonathan S. Levin. 2019. "Does Incorporating Cost-Effectiveness Analysis into Prescribing Decisions Promote Drug Access Equity?" *AMA Journal of Ethics* 21(8):679–85.

Edwards, Ward, and F. Hutton Barron. 1994. "Smarts and Smarter: Improved Simple Methods for Multiattribute Utility Measurement." *Organizational Behavior and Human Decision Processes* 60(3):306–25.

Edwards, Ward, Ralph F. Miles, and Detlof Von Winterfeldt, eds. 2007. *Advances in Decision Analysis: From Foundations to Applications.* Cambridge, UK: Cambridge University Press.

Gibney, Elizabeth. 2019. "'More Women Are Being Nominated': Nobel Academy Head Discusses Diversity." *Nature*, October 4.

Grofman, Bernard, and Scott L. Feld. 2004. "If You Like the Alternative Vote (Aka the Instant Runoff), Then You Ought to Know About the Coombs Rule." *Electoral Studies* 23(4):641–59.

Guinier, Lani. 1995. *Tyranny of the Majority: Fundamental Fairness in Representative Democracy.* New York: Free Press.

Hägele, Günter, and Friedrich Pukelsheim. 2001. "Llull's Writings on Electoral Systems." *Studia Lulliana* 41:3–38.

Kahneman, Daniel. 2011. *Thinking, Fast and Slow.* New York: Farrar, Strauss and Giroux.

Keeney, Ralph. 2009. *Value-Focused Thinking: Path to Creative Decisionmaking.* Cambridge, MA: Harvard University Press.

Laplace, Pierre Simon. 1820. *Théorie Analytique des Probabilités.* Courcier.

Levine, Michael E., and Charles R. Plott. 1977. "Agenda Influence and Its Implications." *Virginia Law Review* 63(4):561–604.

Llull, Ramon. 1280s. "De Arte Electionis [a System for the Election of Persons]."

Madhavan, Guru, and Charles Phelps. 2018. "Human Factors of Democracy." *The Bridge (National Academy of Engineering)* 48(4):40–44.

Madhavan, Guru, and Charles Phelps. 2019. "What If You Could Vote for President Like You Rate Uber Drivers?" *Behavioral Scientist*, April 2.

Madhavan, Guru, Charles Phelps, and Rino Rappuoli. 2017. "Compare Voting Systems to Improve Them." *Nature* 541(7636):151–53.

Maskin, Eric, and Amartya Sen. 2017. "The Rules of the Game: A New Electoral System." *New York Review of Books*, January 19.

McNeil, Barbara, Stephen Pauker, Harold Sox, and Amos Tversky. 1982. "On the Elicitation of Preferences for Alternative Therapies." *New England Journal of Medicine* 306(21):1259–62. doi:10.1056/NEJM198205273062103.

Moore, Blaine Free. 1919. *The History of Cumulative Voting and Minority Representation in Illinois, 1870–1919* (Vol. 8). Urbana: University of Illinois.

Novak, Stéphanie, and Jon Elster, eds. 2014. *Majority Decisions: Principles and Practices.* Cambridge, UK: Cambridge University Press.

Plott, Charles R., and Michael E. Levine. 1978. "A Model of Agenda Influence on Committee Decisions." *American Economic Review* 68(1):146–60.

Roth, Alvin E. 2015. *Who Gets What—and Why: The New Economics of Matchmaking and Market Design.* Boston: Houghton Mifflin Harcourt.

Saari, Donald. 2001. *Chaotic Elections: A Mathematician Looks at Voting*. Providence, RI: American Mathematical Society.

Saari, Donald G. 2001. *Decisions and Elections: Explaining the Unexpected*. Cambridge, UK: Cambridge University Press.

Sen, Amartya. 2017. *Collective Choice and Social Welfare*. Cambridge, MA: Harvard University Press.

Smith, John H. 1973. "Aggregation of Preferences with Variable Electorate." *Econometrica* 41(6):1027–41.

Taber, George. 2005. *Judgment of Paris*. New York: Scribner.

INDEX

For the benefit of digital users, indexed terms that span two pages (e.g., 52–53) may, on occasion, appear on only one of those pages.

Tables and figures are indicated by t and f following the page number.